India, US and the Nuclear Deal
A Critical Study

India, US and the Nuclear Deal
A Critical Study

Mohammed Badrul Alam
Muzaffar Ahemad Ganaie

Ess Ess Publications
New Delhi

India, US and the Nuclear Deal : A Critical Study

ISBN: 978-81-7000-739-5

₹ 1200/-

First Published 2015

Published by:
Ess Ess Publications
4831/24, Ansari Road,
Darya Ganj,
New Delhi-110 002.
INDIA
Phones: 23260807, 41563444
Fax: 41563334
E-mail: info@essessreference.com
www.essessreference.com

Cover Design by *Patch Creative Unit*

Printed and bound in India at Salasar Imaging Systems.

Contents

Acknowledgements

For Mohammed Badrul Alam, this book has its inspiration from teaching a wide range of students in India, USA and Japan over the past three decades. As for inculcation to American Studies discourse, it commenced long time ago in early 1980s in Jawaharlal Nehru University, New Delhi where Mohammed Badrul Alam earned his M.Phil degree in American Studies at the School of International Studies. His professors, Dr. B.K.Shrivastava, Dr. M.S.Venkatramani and Dr. R.P.Kaushik, taught him aspects and various nuances of American Studies and its implications. His long stint working at the American Center library and ICWA Library in New Delhi opened up his eyes to the scope and breadth of literature on almost all important facets of American Studies including strategic discourse.

In the United States, where Dr. Alam attended Cornell University in Ithaca, New York, and earned his M.A and Ph. D degrees in International Studies, many high profile professors such as Prof.Mary Katzenstein, Prof.Milton Esman, Prof.Richard Polenberg, Prof Water LaFeber, instilled in him the various intricacies and complexities of issues related to US-India relations.

In Japan, where he has been teaching at Miyazaki International University in Southern Japan from 1998 to

2006, a number of colleagues such as Dr. Kate Greenfield, Dr. Getachew Felleke, Amanda Bradley, Dr.Jeong Pyong-Hong, Dr.Mika Mervio and Dr.Yasu Izumikawa, helped Dr.Alam in shaping his ideas and in bringing in the right focus on complex dynamics of Indo-US relations.

At Jamia Millia Islamia University, Delhi, where Dr. Alam is presently a Professor of Political Science and its Head, all of his colleagues were of assistance in one way or the other. A number of his colleagues in the Department of Political Science including Prof.Rumki Basu, Prof.Furqan Ahmad, and Prof.Shahid Ashraf, from the Economics Department and Jamia Millia's Registrar as well as Prof.Mujtaba Khan, Director, UGC-Academic Staff College at Jamia Millia Islamia, were great source of inspiration and motivation. A number of doctoral research scholars including Reyaz Ahmad Ganie, Akhilesh Tiwari, Digvijay Singh Sekhawat, Sasrish Sebastian, Kuldeep Singh, Amit Gupta, were of immense help in collecting the source materials.

The greatest gratitude is owed to Dr.Alam's mother, Mrs. Hedatun Nisa and brothers: Dr. Mohammed Shamsul Alam and Mohammed Aminul Alam, and sisters: Dr. (Mrs.) Swaleha Khatun, Mrs. Rahima Khatun, Mrs. Sofia Khatun, his wife: Mrs. Gahartaj Rokeya and the children: Saira Alam, Zubair Alam, Sadaf Alam and Munawar Alam. Although quite demanding and at times frustrating, everyone allowed him the time and necessary emotional support he needed to complete his work. Much of Dr.Alam's success is due to their love and affection at all times.

Muzaffar Ahemad Ganaie is deeply grateful to his supervisor, Prof. Mohammed Badrul Alam, at the Department of Political Science, Jamia Millia Islamia who immensely supported him during his entire Ph. D. programme. This work would not have been possible without his support and cooperation. He would always remain indebted to Prof.Mohammed Badrul Alam for his enduring support and cooperation. Muzaffar is also grateful to the faculty members of the Department of Political Science at Jamia Millia Islamia for their constant support, cooperation and encouragement. His special thanks also go to administration and high dignitaries of Jamia Millia Islamia for providing proper academic ambience and the great, first rate staff members of Dr. Zakir Hussain Library at Jamia Millia Islamia, in providing relevant and very useful data and resource materials for this work. Indian Council of World Affairs, American Center Library and Nehru Memorial Library also provided Muzaffar Ahemad Ganaie all the relevant information and data in completion of this study that formed the core of his doctoral work. This book is based on the Ph.D. Thesis. Muzaffar Ahemad Ganaie has submitted to Jamia Millia Islamia, New Delhi, for the award of the degree of the Doctor of Philosophy. Finally, Muzaffar Ahemad Ganaie is highly indebted to his family members who provided him valuable support during the course of his research. He will always remain grateful to them.

A special word of thanks to Sumit Sethi, New Delhi, India, for showing genuine interest in the manuscript and for its subsequent publication.

Prof.Mohammed Badrul Alam

Muzaffar Ahemad Ganaie

Preface

Indo-US Civil Nuclear Deal also known as India-US Civil Nuclear Cooperation Agreement or 123 Agreement — named after section 123 of US Atomic Energy Act (AEC) of 1954 which makes it obligatory for US to sign an agreement for cooperation as a prerequisite for the transfer of nuclear material, equipment and technology from US to other states — is a bilateral Agreement between India and the United States (US) governing cooperation between two sides in the field of nuclear energy. The framework of this Agreement was laid down by US President George W. Bush and Indian Prime Minister Dr. Manmohan Singh on 18 July, 2005, by issuing a joint statement. In the Joint communiqué, both the leaders expressed their resolve to undertake certain reciprocal measures to realise the potential of full cooperation in the field of civil nuclear energy. US President George W. Bush noted that "he will work to achieve full nuclear cooperation with India" and for this, he would seek an "agreement from Congress to adjust US laws and policies" and "will work with friends and allies to adjust international regimes to enable full civil nuclear cooperation and trade with India."[1] Bush also promised to facilitate India's

1 "Indo-US Joint Statement." Available at http://www.hindu.com/thehindu/nic/indousjoint.htm, (accessed on 12 April 2014)

participation in ambitious international projects ITER and Generation IV International Forum. On his part, Indian Prime Minister stated that as a responsible state, India would also undertake certain reciprocal measures. These would include separating India's civilian and military nuclear facilities in a phased manner, negotiating a Safeguards Agreement with International Atomic Energy Agency (IAEA), continuing her unilateral moratorium on nuclear testing, working towards the conclusion of Fissile Material Cut off Treaty (FMCT) and exercising a strict control on the transfer of sensitive nuclear technology.

The announcement of Civil Nuclear Cooperation Initiative was hailed as a great diplomatic victory in India because it was after a gap of more than thirty years that US had shown keen interest to cooperate with India in the field of civil nuclear energy without interfering in her strategic nuclear programme. The nuclear cooperation between the two sides after a smooth beginning in 1950s had come to an end in 1974 when India detonated a nuclear device at Pokhran making US to terminate nuclear cooperation with India. Since 1974, two sides remained at loggerheads for decades over "nuclear issue" which acted as a major irritant in bilateral relationship. The US policy to look at India through the prism of nuclear proliferation with emphasis on to "cap, rollback and ultimately eliminate" her nuclear weapons programme greatly hampered the process of rapprochement. New realities and challenges of the post Cold War era, however, motivated US to reverse her decades old non-proliferation policy vis-à-vis India and to engage her as a junior partner in global affairs. In this direction, a major and path-breaking initiative undertaken by Bush administration was the Civil Nuclear Cooperation Initiative. With this initiative, George Bush not only overturned the decades old US non-proliferation policy but also removed the biggest obstacle in the bilateral relationship.

Indo-US Civil Nuclear Deal was a very complex deal, as it had to go through several complex stages— separation of India's military and civilian nuclear facilities, amendment of US domestic law, the conclusion of Safeguards Agreement between India and IAEA, and adjustment of the guidelines of Nuclear Suppliers Group (NSG)— to see the light of the day. Towards the operationalisation of the Agreement, the first step undertaken by India was negotiation of "separation plan"— entailing the separation of the civilian and military aspects of her nuclear programme — with US. Subsequently Bush administration enacted "Henry J. Hyde United States-India Peaceful Atomic Energy Act of 2006" to waive certain requirements of US Atomic Energy Act of 1954 to allow US to engage in civil nuclear cooperation with India. The passage of the Act, however, kicked off a political storm in India. The Act contains certain provisions which in India have been viewed as contrary to the July 18 understanding. The Act set in motion a hot debate in India regarding the linkage between the Act and 123 Agreement and its possible implications for India's independent foreign policy and strategic nuclear weapons programme. Thereafter, 123 Agreement was negotiated delineating the necessary responsibilities and conditions of nuclear cooperation between two sides. Other requirements like the conclusion of India-IAEA Safeguards Agreement and adjustment of NSG guidelines were fulfilled in 2008. The Agreement was finally inked by US Secretary of State Condoleezza Rice and her Indian counterpart Pranab Mukherjee on 10 October 2008.

Besides being a complex deal, the Indo-US Civil Nuclear Deal has also turned out to be one of the most controversial deals signed by India in post independence era. The Agreement went into trouble since the day it was announced because of vehement opposition and wide criticism within the parties (India and the US) and outside. In India, the Agreement faced an upfront opposition from certain political outfits and some sections of nuclear scientific establishment and strategic community. The deal was criticised on three main grounds:

- First, the deal would undermine India's strategic nuclear weapons programme.
- Second, the deal would erode the autonomy of India's research and development in the field of nuclear energy.
- Third, the deal would compromise India's independent foreign policy.

Another important issue linked with the Civil Nuclear Cooperation Initiative and which has attracted the attention of analysts and commentators is the role of nuclear energy in meeting India's long term energy requirements. In India, the votaries of the Agreement view nuclear energy as a future source of energy; and the nuclear deal, with its promise of providing nuclear fuel and technology to India, is being seen as a great initiative to boost country's nuclear power generating capacity, which is currently stuck at less than three per cent of total generating capacity. In a nutshell, those who supported the deal were of the view that this could be a major contributor to energy security of India, will provide 20,000 MW of electricity by 2020, open up opportunities for further nuclear cooperation and similar arrangements with other countries, will help India enter into the global mainstream of nuclear power technology, will provide for full scale civil nuclear energy cooperation as well as spur high-end manufacturing in this arena.

A number of observers and analysts have, however, challenged the role of nuclear energy in meeting India's long term energy needs because of the challenges in terms of cost and safety associated with the development of nuclear energy.

In US, the deal faced a stiff opposition from non-proliferation community. It was argued that the deal would have an adverse impact on the countries that are parties to the Treaty on the Non-Proliferation of Nuclear Weapons (NPT) and have renounced their nuclear ambitions with the promise that the benefits of civil nuclear technology will be the prerogative of NPT signatory countries only. From the deal, these states would draw a message that developing nuclear weapons would bring international criticism and sanctions for a limited period, ultimately the rewards would come and a state would be accommodated into global nuclear order as has emerged in the case of India. Secondly, the deal would set a precedent for other states to cut similar deals with their strategic partners and this would greatly undermine the global non-proliferation regime.

In addition to the domestic and global ramifications, another serious issue linked with the initiative is its implications for South Asian region where India is experiencing strained relations with two important stakeholders of the region, Pakistan and China. Strategic analysts have argued that the deal will indirectly enhance India's nuclear weapons capability and this will prompt Pakistan and China to take counter measures. This could trigger an arms race in the region leading to the destabilisation of the entire South Asian region.

This study attempts to provide an in depth analysis of all these issues. Keeping in view the nature of the study, a qualitative approach of research has been applied. In order to provide a deep and thorough analysis of the issues pertaining to the study all kinds of accessible sources like books, journals, newspapers, joint statements, official documents, internet have been consulted.

The present study is divided into five substantive chapters. Chapter one provides a broad overview of India's nuclear programme and the developments in Indian nuclear discourse since its independence. It also describes how and why India became a nuclear weapon state and provides a deep insight into the broad repercussions of her overt weaponisation. It also looks at different aspects of India's nuclear doctrine and the debate surrounding its core principles. Chapter two deals with the genesis and development of India-US Civil Nuclear Cooperation Agreement. It also discusses the motivating factors behind the deal and the controversies surrounding it. Chapter three provides a deep analysis of the process of negotiations at IAEA and NSG and examines the ramifications of the outcome of the negotiations for India. Chapter four analyses the domestic, regional and global ramifications of the Agreement and chapter five puts forward the main findings of the study.

Abbreviations

ABHWR	Advanced Boiling Heavy Water Reactor.
AEA	Atomic Energy Act.
AEC	Atomic Energy Commission.
AEET	Atomic Energy Establishment, Trombay.
AERC	Atomic Energy Research Committee.
AHWR	Advanced Heavy Water Reactor.
AICC	All India Congress Committee.
BARC	Bhabha Atomic Research Center.
BBC	British Broadcasting Corporation.
BCM	Billion Cubic Meters.
BHEL	Bharat Heavy Electrical Limited.
BJP	Bharatiya Janata Party.
BOG	Board of Governors.
BWR	Boiling Water Reactor.
CAGR	Cumulative Average Growth Rate.
CBM	Confidence Building Measure.
CCS	Cabinet Committee on Security.
CD	Conference on Disarmament.
CEA	Central Electricity Authority.
CIL	Coal India Limited.

CIR	Canadian Indian Reactor.
CIRUS	Canadian Indian Reactor, US.
CPI (M)	Communist Party of India (Marxist).
CPI	Communist Party of India.
CSA	Comprehensive Safeguards Agreement.
CSI	Container Security Initiative.
DAE	Department of Atomic Energy.
DPG	Defence Planning Guidance.
DRDO	Defence Research Development Organisation.
ENDC	Eighteen Nation Disarmament Committee.
ENR	Enrichment and Reprocessing.
EPR	European Pressurised Reactor.
EREC	European Renewable Energy Council.
FBR	Fast Breeder Reactor.
FBTR	Fast Breeder Test Reactor.
FMCT	Fissile Material Cut off Treaty.
GDP	Gross Domestic Product.
GW	Gigawatt.
IAEA	International Atomic Energy Agency.
IAEC	Indian Atomic Energy Commission.
IEA	International Energy Agency.
IEP	Integrated Energy Policy.
IPI	Iran-Pakistan-India.
ISRO	Indian Space Research Organisation.
ISSA	India Specific Safeguards Agreement.
KT	Kiloton.
LWR	Light Water Reactor.
MAP	Model Additional Protocol.

MIT	Massachusetts Institute of Technology.
MNRE	Ministry of New and Renewable Energy.
MOU	Memorandum of Understanding.
MP	Member of Parliament.
MT	Million Ton.
MW	Megawatt.
NCA	Nuclear Command Authority.
NCMP	National Common Minimum Programme.
NNWS	Non-Nuclear Weapon State.
NPCIL	Nuclear Power Corporation of India Limited.
NPT	Nuclear Non-Proliferation Treaty.
NSA	National Security Advisor.
NSAB	National Security Advisory Board.
NSG	Nuclear Suppliers Group.
NSSP	Next Steps in Strategic partnership.
PFBR	Prototype Fast Breeder Reactor.
PHWR	Pressurised Heavy Water Reactor.
PM	Prime Minister.
PNE	Peaceful Nuclear Explosion.
PSI	Proliferation Security Initiative
SASO	Southern Arizona Seismic Observatory.
SFC	Strategic Forces Command.
SLBM	Submarine Launched Ballistic Missile.
TAPS	Tarapur Atomic Power Station.
TIFR	Tata Institute of Fundamental Research.
TN	Thermonuclear.
TOE	Tonnes of Oil Equivalent.
UK	United Kingdom.
UN	United Nations.

UNSC	United Nations Security Council.
UPA	United Progressive Alliance.
USSR	Union of Soviet Socialist Republics.
US	Unites States.
VOA	Voluntary Offer Agreement.
WHO	World Health Organisation.

1

India's Nuclear Policy And Its Context

INTRODUCTION

Indian nuclear programme was conceived in the pre-independence era by a small group of influential scientists who were influenced by the discoveries in the field of atomic energy in the West and persuaded Indian leaders to harness nuclear energy for peaceful purposes. In this endeavour, India received assistance from western countries particularly from the United States (US) and Canada. However, India failed to remain committed to develop nuclear energy for peaceful purposes only. Its quest for nuclear weapons commenced when it faced a humiliating defeat in her border war with China in 1962. China's first nuclear test in 1964 further heightened her security concerns and gave India a purpose to acquire nuclear weapons. After much debate and discussion, India broke her nuclear silence in 1974 by conducting a Peaceful Nuclear Explosion (PNE) at Pokhran. After that, India followed nuclear ambiguity for about twenty-four years but ultimately conducted five nuclear tests in 1998 and overtly declared herself a nuclear power. The test blocked India's access to civil nuclear technology which greatly hampered the development of her nuclear power programme and forced India to develop an indigenous nuclear programme. Today, among the developing countries India has a unique nuclear power programme but it has failed to make a substantial contribution to country's power production even after more than forty years of commercial operation.

This chapter traces the developments in Indian nuclear discourse since the beginning of India's nuclear programme up to 1998 nuclear tests. It examines the motivational factors behind India's overt weaponisation and looks at different dimensions of India's nuclear doctrine. The chapter also examines the implications of India's weaponisation for her civil nuclear power programme and analyses its current status and future challenges.

INDIA'S NUCLEAR PROGRAMME: HISTORICAL OVERVIEW

The genesis of India's nuclear programme precedes India's independence. It was kicked off in 1945 with the establishment of Tata Institute of Fundamental Research (TIFR) and subsequently with the creation of Atomic Energy Research Committee (AERC) in 1946. The credit for laying the foundations of India's nuclear programme goes to Dr. Homi Jehangir Bhabha because it was on his insistence that Tata Institute of Fundamental Research, first nuclear research institute of its kind was established in India on 19 December 1945 and Bhabha was made its first Director. In April 1948, Atomic Energy Act (AEA) was enacted which paved the way for the establishment Indian Atomic Energy Commission (IAEC) in 1948 that replaced

AERC and Bhabha was made its first Chairman.

On 3rd January 1954, IAEC decided to setup a new facility—Atomic Energy Establishment, Trombay (AEET) which was later on renamed as Bhabha Atomic Research Centre (BARC) by Prime Minister Indira Gandhi in memory of Bhabha. In order to stimulate nuclear research and development, Department of Atomic Energy (DAE) was created on 3 August 1954 and Bhabha was made its Secretary. Thereafter, with the support of Nehru, India's first Prime Minster who has a personal interest in nuclear energy subsequent developments took place. The atomic energy budget increased twelve fold between 1954 and 1956. In 1954, the construction of India's first nuclear reactor was started with British assistance. It was 1 MW swimming pool- type research reactor and in the same year after tough negotiations of more than one year, Canada agreed to supply 40MW research rector to India, Canada India Reactor, (CIR). The heavy water for the reactor was supplied by US under an agreement signed in 1956 and the reactor was dubbed as Canada-India Reactor, US (CIRUS).[1] The reactor was fuelled with indigenously produced uranium so as to keep control of the plutonium produced there and it was this plutonium which was used in India's first nuclear test called Peaceful Nuclear Explosion (PNE) conducted in1974 though Canada had agreed to supply reactor to India with this understanding that it will be used for peaceful purposes and the agreement signed for the sale of reactor did not mention any type of PNE. However, the agreement was signed at a time when there was no international agency to regulate the export of nuclear material and technology or prevent its diversion which provided India an opportunity to divert the plutonium produced by the reactor.

Right from the beginning, India's nuclear policy was

characterised by duality and ambiguity, which guided her nuclear programme from Nehru to Vajpayee. No Indian Prime Minister in real terms tired to close the option of India developing nuclear weapons. Even Nehru, a great advocate of peaceful uses of nuclear energy did not close the doors for India to develop nuclear weapons as he said in Indian Parliament:

> *"...we must develop this atomic energy quite apart from war—indeed I think we must develop it for the purpose of using it for peaceful purposes. It is in that hope we should develop it. Of course if [we] are compelled as a nation to use it for other purposes no pious sentiments of any of us will stop the nation for using [it] that way."*[2]

Publicly, Nehru was against the development of nuclear weapons because he had a fear that it would lead to the militarisation of Indian society and believed that military spending is a necessary evil but at the same time he did not tie the hands of Indian nuclear scientific establishment. He gave Bhabha a freehand in the development of Indian nuclear infrastructure and Bhabha also worked relentlessly to achieve mastery in all aspects of nuclear fuel cycle.[3] Summarising the views of Nehru on India's nuclear policy Lorne Kavic writes: "Nehru stated his hope that India would develop atomic power for peaceful purposes but warned that, so long as world was constituted as it was, every country would have to develop and use the latest scientific devices for its protection."[4] Nehru's willingness to keep the option of developing nuclear weapons weapon does not mean that he was in favour a nuclear bomb, in fact he was against it but he knew the political value of keeping the option open.[5]

In 1958, India's relations with China strained over border dispute and a severe blow to India came when

China announced that it would soon develop nuclear weapons. This announcement had a strong impact on India's political leadership. In order to allay the concerns of India's political establishment, Nehru declared in Parliament that India faced no threat from China and India is far ahead of China in nuclear research and development and instead of taking counter measures to contain China's threat, Nehru called upon international community to take necessary steps towards the elimination of nuclear weapons. In a statement made in the Indian Parliament, he stated that "If nothing effective is done in regard to disarmament in the course of next three or four years. It may perhaps become too late to deal with it; it may become almost impossible to control the situation."[6] In 1962, war broke out between India and China. In this short but intense border war unprepared and ill-equipped India got a humiliating defeat from China which weakened the political ideology of Nehru and his influence. After the defeat, India significantly diverted its resources towards defence programme. The defence budget for 1963-64 was doubled amounting to twenty-eight percent of the national budget compared to fifteen-seventeen percent in the previous year.[7] No doubt the border war with China did not cause a major shift in India's nuclear policy but it led to first suggestion in the Parliament that India should also develop nuclear weapons. The first voice came from Ramachandra Bade, a Jan Sangh M. P. (Member of Parliament) from Madhya Pradesh. In his speech in Lok Sabha, lower chamber of Indian Parliament, he said, "Only those who wish to see Russians or Chinese ruling India will oppose the development of nuclear weapons. I beg the Prime Minster to make full use of our research in atomic energy."[8] But the suggestion was turned by Nehru on the grounds that development of nuclear weapons is not in India's interests because of huge costs involved in their development and it would weaken India's cause of global nuclear disarmament. On 18 May 1964, nine days before his death, Nehru reiterated his stand in a television interview and said that: "We are determined not to use weapons for war purposes. We do not make atom bombs. 1 do not think we will."[9]

SHASTRI ERA

After the death of Nehru in 1964, a leadership vacuum was created. Lal Bahadur Shastri, a simple looking Congressman, virtually unknown in international politics took over as Prime Minister of India after the death of Nehru. After assuming the office, Shastri had to face two challenges, first Chinese nuclear test of 1964, which set in motion a nuclear debate in the country and second, 1965 war with Pakistan which further exposed India's security weakness. On 16 October, 1964, China conducted its first nuclear test at Lop Nor. The Chinese test revived the demand for the development of nuclear weapons. This nuclear test created apprehensions among the Indian strategic establishment that:

> "*... without its nuclear weapons and without unleashing the kind of war which would be regarded in the west as the crossing of provocation threshold, China may subject a non-nuclear India to periodic blackmail, weaken its people's spirit and self-confidence and thus achieve without a war its major political and military objectives is Asia.*"[10]

Shastri described the test as a shock and danger to the maintenance of world peace and security but showed great reluctance to revive country's nuclear policy. But there was a sharp reaction to Shastri's 'no policy change' stance and demand for manufacturing nuclear weapons started finding its supports within the Congress party also.

Mushtaq Ahmad, President of Delhi Pradesh Congress Committee was the first Congressman who openly supported the development of nuclear weapons. Several Congress MPs privately approached the Prime Minister to convince him to revive country's nuclear policy.[11] On 7th and 8th November 1964, an All India Congress Committee (AICC) meeting was held and in the meeting, majority of members directly supported manufacturing of nuclear weapons. The year 1964 saw the 'nuclear issue' dominating parliamentary proceedings. The winter session of the Indian parliament provides a fascinating description of country's nuclear discourse and positions adopted by different political parties vis-a-vis nuclearisation. While Praja Socialist Party and Jan Sangh were in favour of weaponisation, the Communist Party was against the development of nuclear weapons. Within Congress and Swatantra parties, opinion was divided.[12] The MPs supporting the cause of nuclear weapons were convinced by the figures given by Bhabha, Chairman of AEC regarding the economics of developing nuclear weapons. In 1964, Bhabha had claimed that producing nuclear weapons is quite cheap. He estimated that it would cost \$21 million to manufacture a stockpile of fifty atom bombs and \$ 31.5 million a stockpile of fifty hydrogen bombs.[13] These estimates however proved inappropriate later on but provided the much needed ground for bomb lobbyists to press their demand. But in spite of huge pressure, PM Shastri refused to change country's nuclear policy because he has his own reservations regarding the development of nuclear weapons. He considered it against the principles of morality to develop nuclear weapons and argued that it would weaken the Indian stand on universal nuclear disarmament and can prove economically disastrous for the country. He also had a fear that it would encourage arms race between India and Pakistan and thus destabilise the region. Besides this, country faced certain technical difficulties like inadequate delivery system, and insufficient fissile material [14] which prevented Shastri to change his 'no weapon' policy. After laying the foundation stone of Gandhi library at Vijayawada on 9 November 1964, Shastri said that "India ought not to make nuclear bombs rather ought to work for their elimination" and on 23rd November 1964, he reiterated same position in Indian Parliament. [15]

In September 1965, India and Pakistan fought their second war on disputed territory of Kashmir. The war renewed the demand for nuclear weapons in India. After the end of war nearly one hundred MPs from different political parties including Congress wrote a letter to Shastri urging him to change his stand on nuclear policy. In this war Pakistan received diplomatic support from both China and US. US stopped aid both to India and Pakistan. China increased troop movements along the border and threatened to open second front against India. It also served an ultimatum to India "to remove construction works in Tibet or face grave consequences."[16] The behaviour of China and US during the course of war greatly surprised Indian leaders, but Shastri remained stuck to his stand. Ultimately, due to immense political pressure from different political parties, Shastri approved the 'Subterranean Nuclear Explosive Project' in Dec. 1965[17] which authorised research in nuclear explosives.

SEARCH FOR NUCLEAR GUARANTEE

In late 1964, debate in India over bomb slackened and focus came on securing nuclear security from nuclear powers. An early reference to such a security was made by Bhabha in early1964 when he proposed that "If any state is to be asked to renounce a possible dependence on nuclear weapons to redress the balance of power against a larger and more powerful state not having nuclear weapons, such as China, its security must be guaranteed by both the major nuclear powers."[18] But India's quest for nuclear security stemmed not from Bhabha's proposal but from statement made by US President Lyndon Johnson after China's first nuclear test. In the statement, President Johnson declared that "the nations that do not seek nuclear weapons can be sure that if they need our strong support against some threat of nuclear blackmail then they will have it."[19] In response to President Johnson's assurance of providing nuclear security to non-nuclear states, when Defence Minister Y.B Chavan was asked that whether India would like to request for such a guarantee, Chavan denied. Initially PM Shastri also showed some reluctance but pressure was mounting on him and he soon changed his position. On 4 December 1964, he stated in a press conference that "It was for nuclear powers to discuss some kind of guarantee which was needed not only by India but by all non-nuclear countries."[20] Shastri was looking for a security umbrella covering not only India but all non-nuclear weapon states. But nuclear powers particularly US was not interested in providing such a security umbrella to non-nuclear states. It was not willing to go beyond the statement of President Johnson which was intended to stabilise the political atmosphere in the region after the China's nuclear test. When Shastri realised that his plan was running into trouble he turned towards United Nations seeking multilateral guarantee for no-nuclear states. But this plan also could not materialize due to lack of consensus among the nuclear powers.

In 1966, China conducted its third nuclear test which brought a subsequent change in Indian position and broad nuclear guarantee under the auspices of UN earlier proposed by India got converted to a very limited type of guarantee which Kosygin had proposed that "Nuclear powers would give an undertaking not to use their weapons against non-nuclear states."[21] To this effect, United Nations General Assembly passed a resolution requesting the Disarmament Committee to consider such a proposal urgently but Indian position soon changed when Indian External Affairs Minister M. C. Chagla declared that India had got a unique position and announced that it hoped a joint guarantee from two superpowers, US and USSR, thereby converting the earlier proposed multilateral guarantee into a bilateral guarantee. In order to explore the possibility of securing such a guarantee from two super powers, Indian Prime Minister Indira Gandhi dispatched a senior civil servant, L.K. Jha as personal envoy to Moscow and Washington. But Jha failed in his mission as two super powers were not willing to go beyond the protection in the event of a nuclear attack on India by a nuclear weapon state.[22]

INDIA AND THE NUCLEAR NON-PROLIFERATION TREATY

The Nuclear-Non-Proliferation Treaty (NPT) was opened for signatures in July 1968 and it came into force in 1970 but India being an original sponsor of the treaty decided to stay outside the treaty because of its discriminatory nature. In 1964, India put forward a proposal in UN General Assembly regarding the elimination of nuclear weapons and on November 19, 1965 UN General Assembly adopted a resolution on a 'treaty to prevent proliferation of nuclear weapons' moved by a bloc by eight non-aligned nations led by India. The resolution detailing the principles of the treaty stated that the treaty should be based on following main principles:

- It should prevent the proliferation of nuclear weapons in any form, directly or indirectly.

- The treaty should lay equal obligations and responsibilities on nuclear and non- nuclear states towards the elimination of nuclear weapons.
- It should be a step towards general and complete disarmament.[23]

But none of these principles were incorporated in the final draft of the treaty which made India not to sign the treaty.

India had three strong reservations against the treaty which shaped her decision to stay outside the treaty. First, India wanted a general and complete disarmament, second, it wanted a security guarantee against emerging Chinese nuclear posture and third, it wanted to retain the right to carry out peaceful nuclear explosions.[24] But as the negotiations progressed all these prospects appeared dim which forced India to stay outside.

To begin with, India made its position clear during the deliberations on the treaty that it wanted a complete elimination of nuclear weapons and in this direction a two stage treaty was proposed by Indian delegate to Eighteen Nation Disarmament Committee (ENDC), V.C. Trevedi. In the first stage, the nuclear states would make a pledge not to transfer their nuclear weapons and technology to non-nuclear states, would reduce their exiting stockpiles of nuclear weapons and would stop further production of nuclear weapons. After the implementation of the first stage, in the second stage, the non- nuclear states would undertake not to acquire nuclear weapons by any means.[25] But treaty offered little on this front by allowing the nuclear states to retain their stockpiles of nuclear weapons and only obligated them to negotiate the issue of disarmament in good faith. After India's refusal to join the treaty, Indian Prime Minister Indira Gandhi stated in Parliament that: "India's refusal to sign the NPT was based on enlightened self-interest and the considerations of national policy...nuclear weapon powers insist on their right to continue to manufacture more nuclear weapons."[26]

The main concern of India was the failure of the NPT to left unheeded India's security concerns emanating from China's nuclear weapons. During the deliberations India voiced strong concerns on China's emerging nuclear posture and warned that its decision to join to treaty would be primarily determined by its security considerations. In 1967, M. C. Chagla, India's External Affairs Minister declared in Parliament that: "India was under continuing menace of a country which had already exploded an atomic bomb and we certainly bear in mind this vital factor while considering India's policy towards NPT."[27] Instead of addressing its 'Chinese security threat', the treaty recognised China as Nuclear Weapon State and legalised its stock of nuclear weapons by allowing the possession of nuclear weapons of those states that had acquired such weapons before 1 January 1967.

As noted earlier that after China's 1964 nuclear test, India had undertaken a nuclear explosive project authorised by Prime Minister Shastri and during the negotiations, it stressed that non-nuclear states should be allowed to carry out peaceful nuclear explosions. India insisted that denying this right to non-nuclear states would amount to denying the "benefits of such technology to developing nations".[28] But India's plea was firmly opposed by US on the grounds that there exists no distinction between a "peaceful" and "non-peaceful" nuclear explosion.[29] However, the treaty in its final form provided that the benefits of peaceful nuclear explosions be made available by nuclear states to non-nuclear states and non-nuclear states can avail such benefits pursuant to an international agreement.[30] But this position was firmly opposed by Mr. Azim Hussain, Indian ambassador to

ENDC in these words:

> *"Mankind everywhere should be free not to share in benefits but also to acquire the knowledge of extracting such benefits by itself and to have freedom to use such knowledge. Since nuclear technology is the technology of future and is likely to become a crucial and potent instrument of economic development and social progress, it would obviously be invidious for the greater part of the world to become wholly dependent on few nuclear weapon states for the knowledge and application of this technology..."*[31]

INDIRA ERA

Indian nuclear discourse during the period of Indira Gandhi who became Prime Minister after the death of Shastri reflects continuity as well as change. When China exploded her first nuclear device in 1964, Indira Gandhi who was then Minister for Information and Broadcasting said in an interview that India has mastered necessary nuclear technology and know-how and can produce a nuclear weapon within a period of eighteen months but India should not develop nuclear weapons and should continue to use nuclear energy for peaceful purposes only.[32] After becoming the Prime Minister, she maintained same position but not for a long time. After assuming the office she put brakes on the 'Subterranean Nuclear Explosive Project' authorised by Shastri and showed no intention to break the status quo. In 1968 she declared in Indian Parliament that, "... we do fell that the events of last twenty years clearly show that the possession of nuclear weapons have not given any military advantage in situations of bitter armed conflicts. ... The choices before us involve... engaging in an arms race with sophisticated nuclear warheads and an effective missile delivery system. Such a course, I do not think would strengthen national security... it may well endanger our internal security..."[33]

Indian nuclear discourse underwent a dramatic change in early 1970s. This was the period when major developments were taking place in Indo-Soviet relations on the one hand and Sino-American relations on the other hand and at the same time crises in Pakistan erupted. On July15, 1971, Sino-American breakthrough was achieved and on July 17, 1971, Henry Kissinger declared to India that "['w]e would be unable to help you against China' in the event of Chinese involvement in war between India and Pakistan." [34] More shocking to India was the use of gun-boat diplomacy by US by moving her nuclear warship to the Bay of Bengal. These developments greatly influenced the thinking of Indian political elite. Commenting on the impact of these developments on Indian strategic thinking, B.S. Gupta writes:

> *"The Chinese bomb ceased to be the main argument for the Indian bomb, perhaps because of China's inability to help Pakistan in the 1971 war...The arguments for the bomb now were that without it India couldn't be expected to the corridors of global power, nor enjoy the status of the dominant regional power; that the bomb... would... compel the United States to change its attitude of hostility or benign neglect."*[35]

No doubt, India won the war, which resulted in the disintegration of Pakistan and formation of a new Country, Bangladesh but war time developments greatly influenced the thinking of Indira Gandhi and in 1972 she made a direct reference to a Peaceful Nuclear Explosion (PNE), stating that research is being carried out in the field of

PNE and AEC is constantly reviving the process made in the field taking into account potential economic and environmental hazards.[36]

On 18 May 1974, India carried out a Peaceful Nuclear Explosion (PNE). The test was conducted at Pokhran desert site and was code named as "Smiling Buddha". After the test, New Delhi was quick to declare that test had no military implications and it was a peaceful nuclear explosion. This was done to avoid international reaction. But nobody was going to buy the Indian argument as the basic technology used for developing a nuclear bomb and a peaceful device is same and there is no evidence which shows that India used the PNE for socio-economic or industrial purposes. In 1997, Raja Ramana admitted in an interview that "Peaceful Nuclear Explosive" (PNE) programme was a cover for a programme to develop nuclear weapons.[37] But more important question than the "nature" of the test is what were the real "motives" behind the test. There is no firm evidence to reveal that why such a decision was taken because Mrs. Gandhi never explained why she approved PNE. However, there is a broad consensus among strategic analysts that such a decision was taken to score political gains. Scot D. Sagan claims that it was directed to address domestic political concerns.[38] In 1973-74, a serious drought had hit the country which gave a serious setback to economy of the country and caused social unrest. Corruption and misgovernance had become routine of the day. J. P. movement was launched in Gujrat and in early 1974, a railway strike was called by George Fernandes. All this caused a setback to the popularity of Mrs. Gandhi and she considered it an opportunity to explode a nuclear device and regain her lost public support. Had it been a security matter then there would have been a lot of debate and discussion on it. The decision was taken by Mrs.

Gandhi with the advice of a small circle of advisors and scientists. Defence Minister was informed of the decision only ten days before the explosion and Foreign Minister just forty-eight hours.[39] Former Indian Defence Secretary K. B. Lall also holds the same argument. In an interview in 1984 he said:

> "'The test' arose not out of defence programme... If it was a defence project, there should have been some discussion. . . I know up to May 1973 that . . . the Chairman of the Chiefs of Staff, the Defence Secretary, the Defence Minister [were not] involved.... [I]t did not arise out of thinking from the Defence Ministry or on security counts."[40]

In 1978, Mrs. Gandhi also admitted in an interview with Rodney Jones that her decision was not motivated by "military considerations."[41] Praful Bidwai, a leading journalist and Achin Vanaik also consider "domestic political considerations" as a motivating factor behind the decision[42] and indeed domestic consequences of the test were rewarding. As per Indian Institute of Public Opinion, the test increased the public support for Gandhi by one-third[43] and she regained her lost public support.

On the basis of these arguments it can be surmised that test was directed to gain political benefits than to address security concerns because Indian security environment had not really worsened in 1974 which could have compelled Indira Gandhi to take such a step. From 1971-1974 India had gained strength due to defeat of Pakistan and by signing a treaty of friendship with USSR. If for the sake of argument, it is accepted that US-China nexus and use of gun-boat diplomacy by US had worsened Indian security environment, then it can be hardly said that it could have been addressed by conducting a PNE.

REACTIONS TO PNE

Indian nuclear explosion of 1974 evoked a sharp response from international community. Virtually all the major powers of the world criticised the Indian decision and imposed sanctions on India. Japan and UK cut off their aid to India while Canada accused India of diverting material from CIRUS reactor and broke off all nuclear cooperation with India.US also terminated her nuclear cooperation with India and put brakes on foreign aid to India. It also passed Nuclear Non-Proliferation Act which tightened control on the transfer of nuclear materials and equipment to non-nuclear weapon states. Pakistan termed the test as a "fateful development" and said that the test had serious implications for Pakistan's internal and external policies.[44] Pakistan's Foreign Minster stated that China will continue to support Pakistan against any aggression and nuclear blackmail.[45]In his speech, Chinese Premier Teng Hsiao-Ping reiterated his support to Pakistan and said that China will standby Pakistan to protect its national sovereignty and territorial integrity and will provide every possible support to the people of Kashmir in their struggle for self determination.[46]

TOWARDS POKHRAN II

After the detonation of a peaceful nuclear device in 1974 India surprised the world after a gap of twenty-four years when it conducted five nuclear tests in 1998. In this long period of twenty-four years India's nuclear discourse did not experience any major breakthrough because of political instability in the country. In 1975, India's nuclear programme slowed down. This was a period when Indira Gandhi's popularity got diminished because of political problems resulting in "Emergency" rule and ultimately her ouster in 1977 general elections. After the defeat of Congress, Janata Party came to power with Morarji Desai as Prime Minister. Morarji Desai was against the development of nuclear weapons and during his tenure Indian nuclear programme came to a halt. However, BARC still functioned autonomously and continued to develop some very sensitive weapons related technologies. But Janata government could not survive for a long time and in 1979, Indira Gandhi returned to power and revived country's nuclear programme. In 1982, she gave approval for another round of tests but soon changed her mind giving a simple but an unconvincing reason that: "I am basically against the weapons of mass destruction."[47] The resumption of nuclear testing was again considered in 1995 during the period of P.V. Narasimha Rao but the decision was declined due to lack of consensus on the issue. Barring these two incidents, India's nuclear policy from 1975-1998 did not witness any major change.

A paradigm shift in India's nuclear policy occurred in 1998 when BJP came to power. BJP was the only political party which had openly advocated that it would reconsider India's nuclear policy after coming to power. This found an expression in party's campaign manifesto called "National Agenda for Governance". It underlined:

> *"To ensure the security, territorial integrity and unity of India, we will take all necessary steps and exercise all available options. Toward the end we will re-evaluate the nuclear policy and exercise the option to induct nuclear weapons."*[48]

It was a long term objective of the party to give Indian defence forces nuclear teeth and even the fear of economic sanctions did not deter it. The tests were conducted on 11 May 1998 at Pokhran. The tests received a mixed domestic response. Left parties criticised the government for reversing India's long-standing policy of nuclear weapons. Sangh Parivar hailed the decision while Congress was divided. Some Congress leaders congratulated the nuclear scientific establishment. Sonia Gandhi said that "real strength lies in restraint, not in the display of Shakti" and Manmohan Singh blamed the government for violating national consensus on the nuclear weapons and said that the tests would push country into nuclear arms race.[49]

After the tests, Indian Home Minister Lal Krishna Advani issued a warning to Pakistan leadership by stating that the tests had tilted strategic balance in the region in India's favour and Pakistan must "join India in the common pursuit of peace and prosperity in the subcontinent...any other course will be futile and costly for Pakistan."[50] This and other inflammatory statements increased the pressure on Nawaz Sharif government which gave a matching response by conducting six nuclear tests on 28th May.

EXPLAINING THE TESTS

The Indian decision to conduct nuclear tests at Pokhran in 1998 gave rise to a very important question: Why did India resume testing after a gap of twenty-four years? Several explanations have been advanced to solve the puzzle. One argument holds that it was the "security threat perception" which motivated Indian decision makers to take such a step. After the tests, Prime Minister Vajpayee wrote a letter to US President Bill Clinton providing a justification for the tests. In the letter, he stated that the Indian tests were motivated by "deteriorating security environment, especially the nuclear environment."[51] He further said that the tests were a response to security threat emanating from China's growing nuclear posture and Pakistan's clandestine nuclear activities.[52] But it is misleading to claim that India faced security threat either from China or Pakistan. The relations between China and India were improving since the visit of Indian Prime Minister Rajiv Gandhi to China in 1988. The border was peaceful and tranquil. There were force reductions along the border from both the sides. "Talks over LAC delineation were slow but satisfactory" and China had modified her Stand vis-a-vis Kashmir.[53] Few days before the tests, Chief of the General Staff of China's People's Liberation Army was a guest of the Chief of the Army Staff of India. If the relations between the two countries had been bad, such a visit would not have taken place.[54] There is a consensus among strategic experts that 1960s was more appropriate time for India to conduct a nuclear test. It was the time when India's strategic environment had deteriorated due to Chinese nuclear test and Indo-Pak war over Kashmir and test would have hardly evoked any serious international response, in fact test would have been conducted with the approval of super powers because both the super powers were experiencing sore relations with China. Contrary to this India decided to conduct the tests at a time when it proved counterproductive for it. The tests not only unleashed an arms race in the region but invited strong sanctions for India. "Security threat" from Pakistan also sounds hallow. No doubt two countries had fought several wars in the past but strategic balance in the region was in India's favour and India was having an edge over Pakistan both in nuclear and conventional weapons capability in spite of Pakistan's Gauri missile test in April 1998.[55] In fact, by conducting five nuclear tests India provoked Pakistan to act back leading to the nuclearisation of the region. So "security threat perception" as a ground for carrying out the tests does not hold any weight. Even after the tests, former Indian Prime Minister I. K. Gujral stated in Parliament that there was no security threat to India's national security [56] either from China or Pakistan.

Second argument holds certain strategic developments nuclear and military build-ups in Pakistan and China, China-Pakistan nexus and US tilt towards Pakistan—as a motivating factor behind the tests but there is no solid evidence which supports this argument. So far as nuclear build-ups in Pakistan are concerned such measures were taken to level the power equation in the region which was in India's favour[57] and not to threaten India's national security. It must be noted that Pakistan set her nuclear programme in the course of weapons development only after Indian nuclear test of 1974 which greatly had altered her security calculus and it can be argued that Pakistan's nuclear tests were the corollary of Indian weaponisation. In fact, Pakistan was much interested in nuclear disarmament and in this direction it put forward certain proposals of "nuclear restraint and regional disarmament" but such proposals were turned down by India.[58] On the one hand, India was showing concerns regarding Pakistan's nuclear ambitions and on the other hand it was expanding her weapons programme. By 1990s, India had increased the number of her nuclear weapons to around twenty and had accumulated about three hundred to four hundred and fifty kilograms of plutonium, sufficient to produce about sixty to ninety nuclear bombs.[59] In addition to this, it had made great strides in improving its delivery system by making necessary modifications in it. A breakthrough in this direction was the successful launch of SSM Prithvi in 1988 followed by IRBM Agni I in 1989.[60] So nuclear or military build-ups in Pakistan were taken to ensure its own security and not to threaten India. It is true that China was involved in providing necessary technology and know-how to Pakistan in her military modernisation programme but such assistance was provided to Pakistan to fulfil certain strategic objectives which had hardly anything to do with

India like limiting US influence in Pakistan, cultivating a moderate and an influential Muslim state in order to link itself to larger Muslim world and to get access to sophisticated technology provided by US to Pakistan. [61]

The nuclear developments in China as a compelling factor behind the Indian tests is unconvincing too because such developments were going on in China even before the genesis of Indian nuclear Programme and were directed more toward US and USSR than to India and if China really posed a security threat to India then one can argue that how India "could live with Chinese bomb for a quarter-century without having one of its own"[62] and if intention was to develop a deterrent against China, how it could have been achieved by conducting only five tests and then declaring moratorium on further tests because by 1998 China had around 400-500 nuclear warheads in her basement.

So far as strategic engagement between US and Pakistan is concerned, such an engagement started during the Cold War era and even continues today but US policy had never been to aid Pakistan in getting a nuclear bomb, it even imposed sanctions on China for supplying nuclear technology and know-how to Pakistan.

Third explanation put forth is discriminatory nuclear order. Before 1998, there existed a broad consensus among Indian strategic elite that global nuclear order was unjust not having due regard for India's growing economic and military clout. After India's PNE in 1974 international pressure was mounting on India's nuclear policy as certain developments were taking place in the non-proliferation front. Not only the control on the supply and transfer of nuclear material and technology to non-nuclear weapon states was tightened by the non-proliferation regime but in 1995, NPT was indefinitely extended despite India's strong reservations. In addition to this, Comprehensive Test Ban Treaty (CTBT) was successfully negotiated and adopted in 1996, despite India's opposition. However, what alarmed India was "Entry into Force" clause of the treaty which required all the forty-four nuclear capable states (including India, Pakistan and Israel) to ratify the treaty before it comes in to force. Indian political elite realised that if treaty came in to force, India will lose the option of conducting nuclear tests and would not be able to develop a credible nuclear deterrent.[63] As New Delhi did not want to miss the train and decided to conduct nuclear tests. This argument holds some weight but cannot be cited as a primary reason. It in fact facilitated the decision and made the cause strong for Indian leaders.

The most important reason was BJP's rise into power which exercised the nuclear option to boost its fortunes. During the election campaign, party made the "nuclear issue" most important election issue to mobilize public opinion. For BJP, nuclear weapons had always been a symbol of greatness and it believed that without these weapons India would not be recognised as a great power. After the tests, Brajesh Mishra (Principal Secretary to Prime Minister) said that "...you can't in today's world be counted for something without going nuclear."[64] But more important reason was to boost the fortunes of a sagging coalition government, as it is a well established fact of Indian politics that the fate of a coalition government depends on the allocation of scarce resources such as ministerial positions than on attaching importance to matters of national concern. As BJP was heading a coalition of seventeen parties and in order to score political gains it decided to conduct the tests.

CONTROVERSY OVER THE TEST

One common thing between Pokhran I and Pokhran

II tests is that both witnessed wide spread controversy. Since Pokhran I was dubbed as Peaceful Nuclear Explosion, without having a deterrent value so its controversy was less fierce as compared to Pokhran II tests. In 1998, India conducted five nuclear tests. On 11, May three tests were carried out, a fission device with yield claimed as 15 kilotons, a thermonuclear (TN) device with yield claimed 45 kilotons and a low yield device with yield 0.2 kilotons. On 13 May, two more low yield devices were exploded. Out of the five tests carried out in May, it is the success of thermonuclear (TN) device that has become a controversial issue. No doubt the success of the test is backed by some heavy weight nuclear scientist like A. P. J. Abdul Kalam, R. Chidambaram, Anil Kakodkar and others. But a clutch of Indian and Western scientists have doubted the success of the device. Among the Indian scientists P. K. Iyengar (former Chairman of India's Atomic Energy Commission) was the first person to raise the doubts on the success of the device. After the test, he met BJP leadership and informed it that "we need more tests to develop a thermonuclear weapon, since the one in May 1998 was not a successful one."[65] P. K. Iyengar's view has been supported by the tall figures of Indian scientific establishment like Dr. Santhanam, DR. Homi Sethna and DR. M.R. Srinivasan.

There are two crucial issues underlying the controversy of device, one, "yield of the device" and second, "nature and shape of the carter". Most of the western scientists put the seismic reading of the TN device between 5.2 to 5.4 mb (mb being the body wave magnitude) and argue that at this reading the yield of the device can be between 25-30 KT. According to B. K. Subbarao, a former captain in Indian Navy, at a seismic reading of 5.2 mb the yield of the device has to be less than 30 KT which means that "either the device failed completely or whatever yield was recorded was entirely from fission device."[66] After India's 1998 explosions, George E. Vander Vink, Director of Planning at the Incorporated Research Institute for Seismology noted that, "From Monday's test [May 11, 1998] we have a seismic signal of about 5.4. We think this corresponds roughly to an explosive yield of about 10 to 25 kilotons."[67] On the basis of data collected from twenty-two seismic centres spread in different parts of the world, Terry C Wallace of the Southern Arizona Seismic Observatory (SASO) claims that May 11 tests produced a combined yield of 15 kilotons.[68] This is a very small yield for a thermonuclear device. Yield of TN device comes in megatons. China's first hydrogen bomb produced a yield of over one megaton. But the Indian scientific establishment involved in 1998 tests puts the seismic reading of the device as 5.47 mb with a yield of 45 kilotons.[69] According to R. Chidambaram, an eminent nuclear scientist who led the DAE team in1998 tests, the reasons for recording of lower seismic data on the world seismic stations was the simultaneous detonation of the three devices which resulted in interference of shock waves and led to the recording of lower values by international seismic stations.[70] But how is it possible that interference in shock waves could have affected only international seismic centres and not the Indian ones. Dr. Anil Kakodkar (former Chairman of AEC) and S. K. Sikka claimed in a research work which appeared in BARC News letter No. 172, May1998 under the title "Some Preliminary Results of May 11-13 1998 Nuclear Detonations at Pokhran" that the tests of 1998 gave a body wave magnitude of 5.2 mb on Richter scale and yield produced was 60 kilotons. However, the figures provided by Kakodkar and Sikka come in conflict with figures provided by R. Chidambaram and R. Ramana. These two scientists have claimed in a paper "Some

Studies on India's Peaceful Nuclear Explosion Experiment" that the test of 1974 gave a body wave magnitude of 5.1 mb and estimated yield was 10-12 kilotons.[71] One fails to understand, how the 1998 tests having a Richter Scale value of 5.2 mb which is slightly higher that the value of 1974 test could have produced yield of 60 kilotons, as claimed by Kakodkar and Sikka.

Second issue underlying the controversy is "nature and shape of the carter". According to K. Santhanam who was a field director at Pokhran-II, the TN device did not produce only low yield but it actually failed because after the explosion its shaft remained totally undamaged and it only created a small depression in the shaft mouth. Had it worked, it would have created a carter of 70 metres in diameter.[72] If TN device is buried very deep then it would create a small or even no carter. But according to Santhanam, the device was buried at a depth of 130 metres and fission device at a depth of 100 metres and two devices were separated by a distance of 1 Km. The fission device produced a yield of 15 kilotons and created a big carter but same did not happen in the case of TN device although difference in depth was only 30 metres.[73] Indian scientists claim that it happened due to huge difference in geology but one fails to understand how a small difference of 30 metres can make such a huge difference. So it can be said that the question mark on the success of main TN device still remains and hence over India's credible nuclear deterrence.

INDIA'S NUCLEAR DOCTRINE

India's nuclear doctrine was officially announced on 4 January 2003 when Cabinet Committee on Security (CCS) issued a statement announcing formulation of nuclear doctrine and Command and Control structures for India. The doctrine released by CCS was a reflection of what Indian Prime Minister Atal Behari Vajpayee had stated in Indian Parliament on 27 May, 1998 that India did not "intend to use these weapons [nuclear weapons] for aggression or for mounting threats against any country; these are weapons of self defence to ensure that India is not subject to nuclear threats or coercion..."[74] He had also stated that India had no intentions to engage in arms race and it would "now observe a voluntary moratorium and refrain from conducting underground nuclear test explosion."[75] Earlier, government had devised a twenty-two member National Security Advisory Board (NSAB) and was entrusted with the task of formulating a draft nuclear doctrine. The doctrine prepared by NSAB was released by Brajesh Mishra, National Security Adviser at a press conference on 17 August, 1999. Since it was prepared by NSAB comprising of non-official strategic experts and analysts so it was not considered an official doctrine. This was reiterated by then Indian Foreign Minister Jaswant Sing after its release while responding to a query by stating that: "It is ...not a policy document of the Government of India."[76] However, the Cabinet Committee on Security in its statement adopted almost all the provisions of draft doctrine prepared by NSAB. It stated:

- Building and maintaining a credible minimum deterrent.
- A posture of "no first use".
- Nuclear retaliation to a first strike will be massive and designed to inflict unacceptable damage. Nuclear retaliatory attack will be authorised by the civilian political leadership.
- In the event of a major attack against India, or Indian forces anywhere with biological or chemical weapons, India will retain the option of retaliating with nuclear weapons.

- Continuance of strict control on the export of nuclear and missile related materials and technologies, participation in Fissile Material Cut off Treaty (FMCT) negotiations and continued observance of the moratorium on nuclear tests.
- Continued commitment to the goal of nuclear free word through global verifiable and non discriminatory nuclear disarmament.[77]

So, there are three basic pillars of India's nuclear doctrine- "no first use", credible minimum deterrent and civilian control— rest of the components of the doctrine are derivatives of these three principles.

NO FIRST USE

"No first use" is an old concept. Its origin lies in no first use treaty on chemical weapons and toxins concluded by the international community in 1925.[78] Currently, only India and China are two nuclear states which subscribe to the "no first use" policy. India's policy of "no first use" means that India will not be the first to use its nuclear weapons against any state and will use its nuclear weapons only if attacked with nuclear or biological and chemical weapons. China like India has also a policy of "no first use" in place since 1964 but unlike India, China has retained the option of using its nuclear weapons against its own territory and does not consider it a violation of "no first use" principle. As China considers Taiwan and Arunachal Pradesh its territories so its "no first use" policy does not apply to these territories. Unlike India, Pakistan has subscribed to "first use" policy as it considers India's "no first use" doctrine as a declaratory policy which can be easily changed if need arises.[79]

India's "no first use" policy became a much debated issue after the release of nuclear doctrine and in fact NSAB had advocated review of this principle before the release of the doctrine. This principle has come under sharp criticism from various quarters of defence establishment and senior military officers have always suggested revocation of this principle. It is often argued that it is not advisable to have "no first use policy" in place when no other nuclear power adheres to such a policy. No doubt China is committed to "no-first use" principle but it does not apply to Indian state of Aruncahal Pradesh which China claims its own territory. "No first use" principle speaks of absorbing the first strike of the adversary but after such a strike will India be left with sufficient warheads and delivery system to inflict unacceptable damage to the adversary in retaliation which the doctrine talks about and what about the post-strike disaster plans. These are the questions which haunt India's defence managers. It is also feared that after the first-strike, India will not be in a position to strike back because there will be immense international pressure on India not to retaliate.[80] To quote an Indian Army Chief, "in a nuclear war [the Indian government] would be able to resist international pressure [to stop an Indian counter-strike] I have my serious doubts." [81] Expressing serious doubts regarding "no first use" policy Admiral Prakash, says that this policy is incompatible with the position adopted by Indian government that nuclear weapons are "mere political instruments." By claiming such a position government has to "persuade, compel or coerce" an adversary, without resorting to any sort of force to act in a way that suits Indian interests and this can be achieved only with the threat of use of nuclear weapons. By following the policy of "no first use", such a threat cannot be created. So this policy of no-first use, "severely undermines our deterrence."[82] Expressing a similar concern, former Indian Air Force Chief Anil Tipnis writes:

"Nuclear threat (and hence deterrence)...

> *cannot be created through a potential; weapons have to be up and standing, particularly so if the other side has that position. Between adversaries there cannot be differential in readiness."*[83]

In the backdrop of these concerns, there have been frequent noises to drop "no first use" policy and to replace it with "pre-emption" or a "launch on warning doctrine". But taking into account India's geostrategic location this policy will not suit India. It is strategically unsound to adopt such a policy as it will require the nuclear arsenal being put in hair-trigger alert and to have an advanced infrastructure of early warning system in place. Such a system in the case of India which is surrounded by two close neighbours, Pakistan and China is unworkable as it will hardly take four to eight minutes for a missile to hit a target either launched from India or Pakistan and this is not a sufficient time for early warning. Further, such a system has not proven effective and can lead to a nuclear war due to false alarms. Between 1972 and 1984 US early warning system showed around 20,000 false warnings.[84] The real concern of India's defense managers is not "no first use" policy but China's nuclear posture and Pakistan's "first use" policy and the demand to drop "no first use" policy is the outcome of these concerns. But it is hard to believe these two countries pose any threat to Indian security and their nuclear polices demand India to drop "no first use" policy. China like India has adopted a minimalist view of deterrence, is committed to "no first use" policy and above all has adopted a non-offensive posture of non-deployed weapons and these common things between the two counties are not only guiding their respective nuclear policies but are working as nuclear CBMs.[85] In the past, both sides have shown a great nuclear restraint and in future we could hardly imagine a situation when China will be compelled to use its nuclear forces against India. So far as Pakistan's policy of "first use" is concerned, an undue importance has been given to it. Pakistan has adopted such a policy only to deter conventional superiority of India. No doubt two countries have experienced several conventional wars in the past but it cannot be denied that both have shown a sense of avoiding the use of nuclear weapons, the Kargil War is best example of it and if nuclear war takes place between the two sides it would leave neither country a victor but both vanquished in the killing fields of radio-active debris which knows no borders and it is hard to imagine that the leaders of the two countries will ignore this fact and will engage in any sort of nuclear exchange.

So it is more logical to remain committed to "no first use" policy at least for two reasons, first such a policy reduces the chances of nuclear war and second, it provides enough time to develop credible minimum deterrent which is a fundamental principle of India's nuclear doctrine.

CREDIBLE MINIMUM DETERRENCE

Credible minimum deterrence as concept means to have a small number of deliverable nuclear weapons which can serve the purpose of deterrence. Minimum deterrence as a concept originated in United States. This type of deterrence was practised by US during the time of Henry Truman when less emphasis was being laid on number, warheads were separated from delivery vehicles and were kept in an unassembled state. But after the acquisition of nuclear weapons by Soviet Union, US policy changed. Besides India and Pakistan, China has also subscribed to this policy. Two western nuclear powers, Britain and France have also adopted minimum deterrence policy. Though these two powers have declared minimum deterrence policies but in true sense their deterrents are not "minimum enough".[86]

For India, nuclear weapons are only political weapons not meant to fight war and their sole purpose is to deter the use and threat of use of nuclear weapons. It is for this reason, India has adopted the concept of "credible minimum deterrence"[87] but what it means in quantative terms remains unspecified because no effort has been made to define the term "minimum" and there is no agreement among strategic analysts and defence managers on what constitutes "minimum". Security analysts put the number between twenty-four to four hundred "survivable" warheads. Since India has opted for "no first use" policy and if deterrence breaks then in order to inflict unacceptable damage to adversary India would need double number of warheads than actually required because half of its nuclear warheads will be destroyed in the first strike launched by adversary in spite of best possible concealment and dispersion measures.[88] The number also varies from adversary to adversary. According to General K. Sundarji, in order to deter China and Pakistan, India needs to hit five value targets in Pakistan and ten in China. Each target would at least require three devices. On the basis of these estimates, Sundarji claims that ninety to one hundred and thirty five is an adequate number to absorb the first strike and still have enough warheads to launch second strike.[89] Regarding the number of warheads required for credible minimum deterrence Jasjit Singh says that two to three dozen of nuclear warheads are sufficient but argues that it is the delivery system which is most important. He asserts that "number" also depends on survivability and Submarine Launched Ballistic Missiles (SLBMs) are very crucial for survivability but India is still far behind to have operational SLBM system.[90] Like Jasjit Singh, Lt. General

Kapur also attaches more importance to survivability than "number" and claims that credible deterrence comes from protecting ones weapons and effective delivery system and for this purpose triad is very important.[91] On the other hand, Bharat Karnad argues that for credible minimum deterrence India needs four hundred to four hundred and fifty nuclear weapons along with thermonuclear and tactical weapons. So, there is no agreement on what constitutes minimum in terms of "number" and there cannot be, Jaswant Singh has rightly said that: "There is no fixity to minimum, there cannot be, because it is a variable, in technology terms, in terms of physical numbers, it is a variable in terms of the launch capacity."[92] The "number" of nuclear weapons to ensure a credible minimum deterrence depends on many factors like level of yield of nuclear weapons, survivability, defence capability to absorb first strike and effectiveness of delivery system. If the nuclear weapons are of higher yield then definitely less number will be required to inflict an unacceptable damage to adversary. Defence capability of an adversary can also have a bearing on number as advanced defence system can render effective delivery vehicles ineffective. Finally, survivability and effective deliver system will definitely demand less number because it is not important how many weapons does a country possess but how many will survive and hit the adversary.

There is a good number of strategic analysts who does not attach much importance to number of weapons and argue that deterrence is not a numbers game. It is an effective delivery system which is very important for deterrence. During the Cold War era, the strength of Soviet nuclear arsenal was only one-seventh of US stockpile but it still served as credible deterrent against US.[93] Rajesh Basrur holds that deterrence does not depend on number of weapons but on the "adversary's perception of risk and certainty of catastrophic consequences" and it was this perception which prevented Russia from launching even a "preventive strike" against a Chinese nuclear facility.[94] Kenneth Waltz, a leading international theorist also does not consider "number" important for credible deterrence. He asserts that a few survivable warheads with effective deliverable system are sufficient to deter the adversary. During the Cuban Missile Crises, US had number advantage over Soviet Union and in the Ussuri river crises, Soviet Union enjoyed such an advantage over China but both were deterred by the second strike capability of their adversaries.[95]

From the arguments put forth by different analysts, it becomes obvious that there is no consensus on what constitutes credible minimum deterrence in terms of number of nuclear warheads, however, a broad consensus exists on the importance of effective delivery system and survivability of nuclear weapons which can ensure credible minimum deterrence and this requirement can be fulfilled by nuclear triad i.e. three nuclear legs comprising of land, air and sea components (air force based bombers, army controlled missiles and navy held submarines). Does India posses all these components to ensure a credible minimum deterrence? is a question which demands an answer. Prithvi and Agni family of missiles constitute the land leg of Indian triad and both have several variants. Prithvi family includes Prithvi I with a range of 150 kilometres, Prithvi II with a range of 250 kilometres and Prithvi III with a range of 350 kilometres.[96] Agni family includes Agni I with a range of 700-800 kilometres. It is a shorter range missile especially designed to strike targets in Pakistan. Agni II, an operational version of Agni I is an intermediate range ballistic missile having a range of more than 2000 kilometres, Agni III (not deployed yet) has a range of 3,500 kilometres and Agni V, an upgraded version of Agni III, is still under development. It is an intercontinental ballistic missile having a range of 5000-6000 kilometres.[97] So all these missiles have a limited range which limits their deterrent value and are more vulnerable to surprise attacks in this modern age of technology.

Jaguars and MIGs comprise the air leg of Indian triad. India has 147 MIG-27s and 88 Jaguars which can deliver nuclear weapons only after modifications. Besides this, India has 150 MIG-21 fighters, 64 MIG-29s and 36 Mirage 2000 which have no nuclear role.[98] All these aircrafts are vulnerable to anti aircraft weapons and cannot penetrate deep into enemy territory to hit a target.

The most reliable component of nuclear triad is the sea leg comprising of Submarine Launched Ballistic Missiles (SLBMs) and this is the weakest part of Indian triad. The submarines have the quality of endurance and survivability which makes them a vital component of nuclear triad. Submarines can remain under water for months which makes their detection most difficult and it is for this reason all the nuclear powers have given much importance to developing and improving their nuclear submarine fleets. Currently, India has ship launched nuclear missiles developed under Danush programme and SLBM Sagarika which has a limited range which limits its deterrent value. However, India is working on longer range SLBMs but it will take it at least one decade or more to have an operational fleet of submarines and Submarine Launched Ballistic Missiles.[99] In addition to this, India is working on anti-ballistic defence system with the assistance of US and Israel and in this regard has conducted several successful tests. India is also working on laser-based anti-missile system which can intercept the missiles soon after their launch. Once put in place, these systems can ensure India an effective second strike capability but have serious regional implications and can trigger arms race in the region.

CIVILIAN CONTROL

Civilian control on the use of nuclear weapons is an important component of January 2003 nuclear doctrine. This control is to be exercised through three-tier Nuclear Command Authority (NCA). NCA comprises of Political Council, Executive Council and Strategic Forces Command (SFC). The Political Council is chaired by Prime Minister of India and it is the body which can authorise the use of nuclear weapons. Its other members include Defence Minister, Finance Minister and Home minister. The Executive council is chaired by National Security Advisor (NSA). This body provides inputs for decision making and executes the directions given to it by Political Council. This body includes senior bureaucrats, the three Service Chiefs, the Intelligence Chiefs, the Chiefs of DAE and DRDO.

SFC is headed by Commander-in-Chief who is of the rank of Air Marshal or its equivalent. This body operationalizes the directives of NCA. It manages and administers all the nuclear forces. The initiation of the process of delivery of nuclear weapons and exact selection of the target is also the responsibility of SFC. But all this is done with the formal approval of NCA. This body comprises of the representative of all the three services and an adequate number of civilian staff from Atomic Energy Commission and missile experts from DRDO.

Though nuclear doctrine places nuclear button under the control of Prime Minister but this does not mean that only PM can authorise the use of nuclear weapons as press release of January 2003 also talks about "alternate chains of command for retaliatory nuclear strikes in all eventualities." But it is not clear who constitute these chains of command. However, it is expected that in the event of a nuclear attack on India eliminating the civilian political leadership and paralyzing the entire communication system, a person at lower level in the command structure can authorise the use of nuclear weapons. [100]

INDIA'S CIVIL NUCLEAR POWER PROGRAMME

Nuclear power generation was on the agenda of Indian scientific establishment even before independence and after independence it was Nehru, who had a personal interest in nuclear energy provided all possible support to the development of nuclear energy for peaceful purposes. While inaugurating India's first nuclear reactor in 1957, Nehru said that:

> "'Atomic Revolution' was like the 'Industrial Revolution'; if India did not develop atomic energy, it would lose again. Either you go ahead with it or you succumb and others go ahead and you fall back and gradually drag yourself along in the trail."[101]

In this endeavour to exploit atom for peaceful purposes, India received assistance from Western countries particularly from United States and Canada. India's commercial nuclear programme went on line in 1969 with the operation of Tarapur Atomic Power Station (TAPS). In 1974, India conducted a PNE which led to end of foreign cooperation and assistance. After the test, India was subjected to international sanctions which left India isolated in international nuclear order and greatly hampered the development of nuclear power generation in the country. In mid 1990s, its power reactors had some of the world's lowest capacity factors, reflecting the technical difficulties of the country's nuclear isolation. However, these sanctions proved a blessing in disguise for India as India was forced to develop her nuclear programme with indigenous resources and to gain mastery over the necessary nuclear technology and know-how.

Today, India is one among the fast growing economies of the world and second largest country in the world with burgeoning energy needs. In order to meet the future demands of energy and keep her engine of economic growth running, India has to exploit every possible source of energy. According to Planning Commission's Integrated Energy Policy document, to maintain a growth rate of 8-10 per cent over next 25 years India has to increase her primary energy supplies by 4 times.[102] In India, per capita power consumption is around 400 kwh/yr which is much lower than the world standards of 2400kwh/yr. According to UN Human Development Index, there exists a strong correlation between the per capita energy consumption and quality of life on the one hand and per capita energy consumption and income of a nation on the other hand. So in order to meet the world standards of per capita consumption of energy, India has to exploit all its domestic sources of energy. But from long term perspective India has limited options. By the end of 2012, total installed capacity of the country was 211766 MW comprising of 141714 MW from thermal, 39416 MW form hydro, 25856 MW from renewable sources and 4780 MW from nuclear power.

Coal is the main-stay of India's power generation but for long term perspective it is not considered a viable option. India's coal reserves are estimated to be around 250 billion tons, with 95 billion tons as proven reserves. According to the estimates of planning commission of India, if coal consumption increases at 5 per cent per annum, then economically extractable coal could peter out in next 25 years.[103] Besides this, burning of coal has

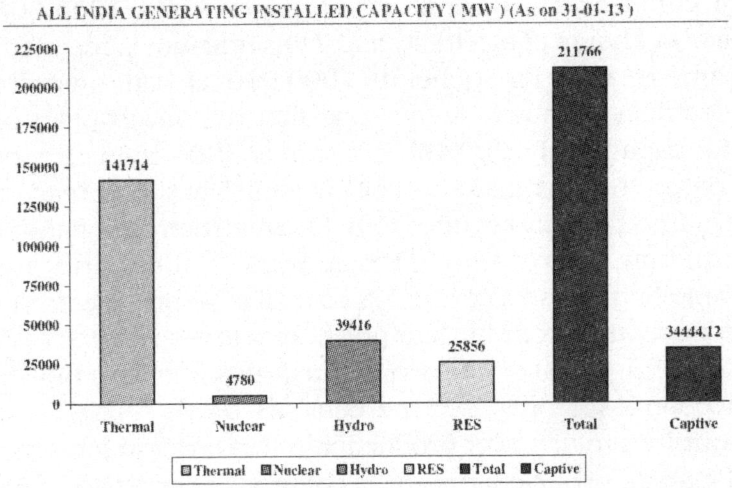

Fig. 1.1

(***Source:*** http://www.cea.nic.in/reports/monthly/executive_rep/jan13/8.pdf)

a very harmful affect on environment as it produces greenhouse gases which lead to global warming and climate change. Hydroelectricity is environmentally clean and India has vast hydro potential of which only a small part has been exploited so far, but problem with this source is that storing of water in reservoirs leads to submerging of forests causing environmental damage and displacement of large sections of population. Harnessing of solar energy is dismissed because solar cells are expensive and solar arrays require large land surfaces. Wind power depends on climate and tidal, geothermal and other non-conventional sources are still at an experimental stage. In this scenario Indian policy makers have chosen the option of nuclear energy to meet future demands of energy. Nuclear energy is a robust source of energy. One ton of uranium produces as much energy as produced by 2 to 3 million tonnes of fossil fuels. Burning of 1 kg of firewood generates 1 kilowatt hour of electricity and burning of 1 kg of coal and oil produce 3 and 4 kilowatt hours of electricity respectively. In comparison, 1 Kg of uranium generates 4000,000 kilowatt hours of electricity and if this uranium is recycled, same quantity can generate 7000,000 kilowatt hours of electricity. Annually an average thermal power plant of the capacity of 1000MWe consumes 2000 train cars of coal or 10 super tankers of oil but a nuclear power reactor of same capacity requires only 12 cubic metres of natural uranium.[104] Presently 31 countries in the world are exploiting this source and 17 per cent of world's electricity is generated from nuclear power reactors spread across different parts of the world. In India, nuclear power accounts for only 2.25 percent (as on 31-1-2013) of country's total power production but in order to increase the share of nuclear energy in country's energy mix India has undertaken a very ambitious project. India's Department of Atomic Energy has set a target to generate 20000 MW of nuclear power by 2020 and 63000 MWe by 2032. [105] DAE is pursuing a three stage nuclear power programme to make best use of country's resources to met bourgeoning demands of energy.

THREE STAGE POWER PROGRAMME

India has modest reserves of uranium but abundant reserves of thorium. Only 1 percent of world's uranium reserves are located in India but country has 30 per cent of world's thorium reserves.[106] So the main objective is to exploit these vast reserves of thorium to achieve self sufficiency in nuclear energy by following a three stage nuclear programme. The three stage programme was introduced by Dr. Homi Bhabha and Dr. N. B Prasad in a paper entitled "Study of the Contribution of Atomic Energy to a Power Programme in India", presented at Second International Conference on the Peaceful Uses of Atomic Energy in 1958.[107] This programme is based on closed fuel cycle in which spent fuel of one stage is reprocessed and used as fuel in the second stage.

First stage : This stage comprises of Pressurised Heavy Water Reactors (PHWRs). These reactors use natural uranium as fuel. In these reactors uranium 235 (U^{235}) which constitutes 0.72 percent of natural uranium undergoes fission and remaining 99.28 percent of uranium 238 (U^{238}) gets converted into Plutonium 239 (Pu^{239}). The spent fuel from these reactors is reprocessed to recover PU^{239} which is used as fuel along with thorium in the second stage. This stage has to serve two objectives, one to produce electricity and second to generate fuel for second stage. Selection of PHWRs for the first stage was made to avoid huge costs required for developing enrichment facilities because these reactors use natural uranium as fuel and thus eliminate the need of sensitive technologies which require huge amount of money.[108]

Second stage : This Stage comprises of Fast Breeder Reactors (FBRs). These reactors use U^{238} and Pu^{239} as fuel. Pu^{239} undergoes fission producing energy as well as Pu^{239} by transmutation of U^{238}. Besides U^{238} a blanket of thorium -232 in the reactor also undergoes transmutation leading to the formation of U^{233} which is used a fuel in the third stage. These reactors produce more fuel than they actually consume.[109] The objective of this stage is to produce power and fuel for the second and the third stage. This stage is in technology demonstration domain.

Third stage : This stage comprises of Advanced Heavy Water Reactors (AHWRs). These reactors use U^{233} as fuel generated from second phase. Besides U^{233} these reactors use a thorium blanket around the U^{233} reactor core which generates more U^{233}, thus producing more and more fuel. The objective of this stage is to provide long-term energy security to country by developing thorium-uranium 233 system that exploits vast thorium reserves of the country. This stage is still in technology development phase.

CURRENT STATUS AND FUTURE PLANS

Currently India operates twenty nuclear power reactors with an installed capacity of 4780 MWe. Among these reactors two are Boiling Water Reactors (BWRs) and rest Pressurised Heavy Water Reactors (PHWRs). Besides these reactors six are under construction which include two1000 MW VVER type reactors being built at Kundankulam, Tamil Nadu, two 700 MW PHWRs at Kakrapar in Gujarat and two 700MW PHWRs at Rawatbtatta in Rajasthan. A 500 MW Prototype Fast Breeder Reactor is also under construction at Kalpakam. This reactor is being built by Bharatiya Nabhikiya Vidyut Nigam Limited (BHAVINI), a government enterprise set up under DAE to focus on Fast Breeder Reactors. Government also has a plan to build four additional 500 MWe FBRs by 2020. In addition to this a 300 MWe Advanced Heavy Water Reactor (AHWR) is being developed at Trombay. According to Department of Atomic Energy Annual Report 2009-2010 a newer version of AHWR that uses lower enriched uranium along with thorium as fuel has been designed. Since these reactors are in experimental stage so India has to wait for decades to see their operationalisation.

In order to achieve the target of 20000 MWe by 2020 set by DAE India has to expand its fleet of nuclear reactors and for this purpose plans are to build 48 reactors by 2023[110] and to this effect a number of agreements are in offing and some have been signed with foreign nuclear vendors. In 2010 a 'roadmap' for building six reactors at Kundankulam by 2017 and four more at Haripur was agreed. In February 2009 Areva signed a Memorandum of Understating (MOU) with Nuclear Power Corporation of India Limited (NPCIL) to build two and later four EPR (European Pressurised Reactors) units at Jaitapur. In March 2009 GE Hitachi Nuclear Energy signed agreements with NPCIL and Bharat Heavy Electricals Limited (BHEL) to build a multi unit nuclear power plant of 1350 MWe using Advanced Boiling Heavy Water Reactors (ABHWRs). United States' Westinghouse and Koreas Electric Power Company are also likely to get contracts for building nuclear power reactors in India[111]. This process will strongly establish the first stage of India's nuclear power programme.

Table 1.1. India's operating nuclear power reactors

S. No.	Plant	Unit	Type	Capacity (MWe)	Date of commercial operation
1.	Tarapur Atomic Power Station(TAPS)	1	BWR	160	28 Oct.1969
2.	Tarapur Atomic Power Station(TAPS)	2	BWR	160	28 Oct.1969
3.	Tarapur Atomic Power Station(TAPS)	3	PHWR	540	18 Aug. 2006
4.	Tarapur Atomic Power Station(TAPS)	4	PHWR	540	12 Sept.2005
5.	Rajasthan Atomic Power Station (RAPS)	1	PHWR	100	16 Dec. 1973
6.	Rajasthan Atomic Power Station (RAPS)	2	PHWR	200	1 April 1981
7.	Rajasthan Atomic Power Station (RAPS)	3	PHWR	220	1 June 2000
8.	Rajasthan Atomic Power Station (RAPS)	4	PHWR	220	23 Dec. 2000
9.	Rajasthan Atomic Power Station (RAPS)	5	PHWR	220	4 Feb. 2010
10.	Rajasthan Atomic Power Station (RAPS)	6	PHWR	220	31 March 2010
11.	Madras Atomic Power Station (MAPS)	1	PHWR	220	27 Jan.1984
12.	Madras Atomic Power Station (MAPS)	2	PHWR	220	21 March1986
13.	Kaiga Atomic Power Station	1	PHWR	220	16 Nov.2000
14.	Kaiga Atomic Power Station	2	PHWR	220	16 March 2000
15.	Kaiga Atomic Power Station	3	PHWR	220	6 May 2000
16.	Kaiga Atomic Power Station	4	PHWR	220	27 Nov. 2010
17.	Narora Atomic Power Station (NAPS)	1	PHWR	220	1 Jan.1991
18.	Narora Atomic Power Station (NAPS)	2	PHWR	220	1 July 1992
19.	Kakrapar Atomic Power Station(KAPS)	1	PHWR	220	6 May, 1993
20	Kakrapar Atomic Power Station(KAPS)	2	PHWR	220	1 Sept.1995

(*Source:* Strategic Analysis, Vol. 35, No.3, May, 2011.)

Table 1.2. Nuclear power projects under construction

S. No	Project	Unit	Capacity (MWe)	Expected commercial operation
1.	Kundankulam Atomic Power Project	1	1000	June, 2011.
2.	Kundankulam Atomic Power Project	2	1000	March, 2012.
3.	Rajasthan Atomic Power Project	7	7000	June, 2016.
4.	Rajasthan Atomic Power Project	8	7000	December, 2016
5.	Kakrapar Atomic Power Project	3	7000	June, 2015
6.	Kakrapar Atomic power Project	4	7000	December, 2015.

(***Source:*** http://www.npcil.nic.in)

FUTURE CHALLENGES

Among various sources of energy, nuclear energy has the highest potential but there are maximum risks involved in the use of this source of energy. There are three key issues involved in the use of nuclear energy: safety, cost and waste management. These three issues have always hindered the development of nuclear energy and have raised serious doubts among the world countries regarding the use of nuclear energy. Nuclear power generation is not considered safe. Past nuclear accidents like Three Mile Island accident, Chernobyl nuclear disaster and Fukushima nuclear accident have brought to light the dangers involved in exploiting nuclear energy for power generation. These accidents have forced many countries to abandon the course of nuclear power development. Fukushima nuclear accident has proved catastrophic for global nuclear industry and after the accident a number of countries have decided to phase out their nuclear power facilities and many have decided not to develop nuclear power at all. Nuclear power industry is now being viewed with fear and suspicion across the world. The industry is witnessing a widespread public opposition throughout the world. India is also facing a stiff opposition from varies social activist groups and local people against the development of nuclear power which has led to delays in the operationalisation and construction of many nuclear projects. There were widespread public protests against Jaitapur power project and Kundankulam power project. The work at two Kundankulam nuclear power projects in Tamil Nadu has come to halt because of stiff opposition posed by the local people who were demanding complete shutdown of the projects because of safety reasons and environmental concerns.

Nuclear power is not considered cost effective because of huge amounts involved in the establishment of nuclear power projects. According to a study conducted by MIT, one megawatt of electricity generated by a nuclear reactor costs $ 2 million and in the case of Light Water Reactors(LWRs) it could be up to $ 3 million.[112] Additional costs are also involved in the management of waste and this is much higher than the cost involved in the thermal or hydropower generation. As India plans to generate 20000 MWs of nuclear power by 2020, it has to spend around $ 65 billion between 2010 and 2020 in power plants, in addition to this huge amount will also be required to feed her operational nuclear fleet.[113]

Waste management is also a serious issue linked with the development of nuclear energy. Nuclear reactors produce radioactive waste which has serious health and environmental implications. This waste remains active for centuries and there is no well established technique to dispose it off. One way out is to dump it in a stable geological repository but only one such repository exits in the world, in US and that too manages military waste only. As India has opted for a closed fuel cycle in which spent fuel is reprocessed and further used but still in this process a significant amount of radioactive waste is produced. In 2001 it was estimated that about $5000m^3$ of high level waste (about the size of two Olympic size swimming pools) has been generated in India. DAE claims that it will manage the waste in a deep geological repository but admits that it is a major challenge.[114]

Operation of the first stage of nuclear programme at a higher level is very important for the success of second and third stages and to meet growing demand of electricity. This requires an adequate and uninterrupted supply of uranium. But in India production of uranium, is not adequate to make the existing plants to work at higher levels. In order to make the PHWRs to function at 85% capacity factor, 605 tons of Uo_2 will be required annually. This figure will reach to 1600 tons of UO_2 if the installed capacity reaches to 10,060 MWe at the capacity factor of 85 per cent and if this requirement is met India's estimated indigenous reserves of uranium would peter out by 2047.[115]

For the success of second and third stages, reprocessing of spent fuel is very important. Currently, India has four reprocessing facilities, two at Tarapur, one at Trombay and one at Kalpakam. The designed capacity of these facilities is sufficient to meet the requirements of 500 MW FBR over next ten years. As India plans to build four such reactors in future so it has to build a number of high capacity reprocessing facilities to meet their requirements but this will be a major challenge as Japan has to wait for two decades to see its Rokkasho Reprocessing plant working.[116]

Further, about 90 per cent of India's projected power capacity is to come from second and third stages but these stages are still in technological development, demonstration stage and it may take decades to see their commercial operation. The Prototype Fast Breeder reactor

(PFBR) was scheduled to come on line in 2010 but this has not happened. The project was taken up in 1980s, and in this sense it can be said that project is running from last 30 years but has not seen light of the day yet. Furthermore, these stages use technology that is not used commercially anywhere in the world because such a technology is considered unsafe. Many countries have used and given up this technology. No country in the world has tried to build a thorium reactor of the kind as India is planning to build for its third stage nuclear programme[117] because of the cost and safety issues.

CONCLUSION

A critical analysis of India's nuclear policy reveals that it was "security threat perception" coupled with domestic pressure which made India to divert its resources towards the development of nuclear weapons but the final decision to "go nuclear" was not a response to any security threat but to score political gains. Indian leaders conflated their personnel interests with national interests and personal security with national security to justify their decision. Further, the tests did not serve India any major goals as after the tests country was not only subjected to nuclear isolation but it lost strategic advantage which it used to enjoy over Pakistan. After the Indian tests, a new strategic balance was created because of Pakistan's nuclear tests which over-shadowed India's conventional superiority over Pakistan. The tests also failed to provide India a "credible deterrent" because of the failure of main thermonuclear device. However, the biggest causality of India's nuclearisation was her civil nuclear power programme as right after the test of 1974, India was subjected to international sanctions which severely affected her nuclear power programme. However, no such sanctions exist today and country has undertaken an ambitious project to expand its nuclear power generation capacity but is facing stiff public opposition due to safety concerns and it can be said that coal, which is the main contributor to country's energy mix will dominate the energy scenario of the country for a long time.

REFERENCES

1 "India's Nuclear Weapons Program: The Beginning: 1944-1960." Available at http://www.nuclearweaponarchieve.org/India/Indiaorigin.htm, (accessed on 5 July 2011).

2 Quoted in S. K. Mishra (ed.), 2006, *India's Nuclear Policy, Disarmament and International Security,* New Delhi: Radha Publications, p. 11.

3 Sumit Ganguly, Spring 1999, "India's Path To Pokhran II: The Prospects and Sources of New Delhi's Nuclear Weapons Programme," *International Security*, vol. 23, no. 4, pp. 150-151

4 Quoted in Ashok Kapur, November, 1974, "India's Nuclear Presence," *The World Today*, vol. 30, p.462.

5 *Ibid.*

6 Quoted in Bhumitra Chakma, February 2005, "Towards Pokhran II: Explaining India's Nuclearisation Process," *Modern Asian Studies,* vol. 39, no. 1, p. 194.

7 Dilip H. Mohite and Sumita Pawar, 2001, *Pokhran Vs Chhagai Hills: The Nuclear Dilemma for India and Pakistan*, Delhi: Kalinga Publications, pp. 36-37.

8 Quoted in Shyam Bhatia, 1979, *India's Nuclear Bomb*, Ghaziabad: Vikas Publishing House, pp. 108-109.

9 Quoted in Sumit Ganguly, "India's Path to Pokhran II: The Prospects and Sources of New Delhi's Nuclear Weapons Programme," *op. cit.*, p. 153.

10 Sisir Gupta quoted in *Ibid.*, p.152

11 Shyam Bhatia, "India's Nuclear Bomb," *op. cit.*, p. 110.

12 *Ibid.*, p. 111.

13 "India's Nuclear Weapons Programme: On to Weapons Development: 1960-1967." Available at http://www.nuclear weaponarchive.org/India/indiaweapons development.html, (accessed on 7 July 2011.)

14 Shyam Bhatia, "India's Nuclear Bomb," *op. cit.*, pp. 126-129.

15 *Ibid.*, p. 120.

16 George Perkovich, 2000, *India's Nuclear Bomb: The Impact on Global Proliferation*, New Delhi: Oxford University Press, p. 109.

17 Col. Ravi Nanda, 1998, *Strategic Compulsions of Nuclear India*, New Delhi: Lancer Books, p.30.

18 Quoted in George Perkovich, *India's Nuclear Bomb: The impact on Global Proliferation*, *op. cit.*, p. 61.

19 Quoted in S.K. Mishra (ed.), *India's Nuclear Policy, Disarmament and International Security*, *op. cit.*, p. 14.

20 Quoted in A. G. Noorani, July 1967, "India's Quest for Nuclear Guarantee," *Asian survey*, vol. 7, no. 7, pp. 491.

21 A. G. Noorani, "India's Quest for Nuclear Guarantee," *op. cit.*, p. 496.

22 For further details see A. G. Noorani, "India's Quest for Nuclear Guarantee," *ibid.*, pp. 496-499.

23 N. D. Jayaprakash, February 12-18, 2000, "Nuclear Disarmament and India," *Economic and Political weekly*, vol. 35, no. 7, p. 526.

24 Bhumitra Chakma, "Towards Pokhran II: Explaining India's Nuclearisation Process," *op. cit.*, p. 208.

25 James E. Dougherty, January 1966, "The Non-Proliferation Treaty," *Russian Review*, vol.25, no. 1, pp. 21-22.

26 Quoted in Leonard Weiss, March 2010, "India and the NPT," *Strategic Analysis*, vol. 34, no. 2, p. 262.

27 Quoted in Bhumitra Chakma, "Towards Pokhran II: Explaining India's Nuclearisation Process," *op. cit.*, p. 208.

28 *Ibid.*, p.209.

29 Summit Ganguly, "India's Path to Pokhran II: The Prospects and Sources of New Delhi's Nuclear Weapons Programme," *op. cit.*, P. 155.

30 Edwin Brown Firmage, October 1969, "The Treaty On The Non-proliferation of Nuclear weapons," *The American Journal of International Law*, vol. 63, no. 4, pp.728-729.

31 Quoted in R. R. Subramanian, 1975, "Nuclear India and the NPT: Prospects for Future," *Instant Research on Peace and Violence*, vol. 5, no. 1, p. 37.

32 P. M. Kamath, 2009, *India's Policy of No first Use of Nuclear Weapons*, New Delhi: Anamika Publishers , p. 54.

33 Quoted in Praful Bidwai and Achin Vanaik, 1999, *South Asia on A Short Fuse*, New Delhi: Oxford University Press, p. 58.

34 George Perkovich, *India's Nuclear Bomb: The Impact on Global Proliferation, op. cit.*, p. 163.

35 Quoted in "India's Nuclear weapons Programme: India's First Bomb: 1967-1974." Available at http://www.nuclearweapon archive.org/India/Indiasfirstbomb.html/, (accessed on 7 July 2011).

36 Shrikant Paranjpe, Summer 1986, "India's Nuclear Status: A Policy Note," *Asian Affairs*, vol. 13, no. 2, p. 57.

37 "India's Nuclear Weapons Programme: India's First Bomb: 1967-1974," *op.cit.*

38 Scot D. Sagan, Winter 1996-97, "Why Do States Build Nuclear Weapons: Three Models in Search of a Bomb," International Security, vo. 21, no. 3, pp. 67-68.

39 *Ibid.*, p. 67.

40 Quoted in Perkovich, *India's Nuclear Bomb: The Impact on Global Proliferation, op. cit*, pp. 177.

41 *Ibid.*

42 Praful Bidwai and Achin Vanaik, 2000, *New Nukes: India, Pakistan and Global Disarmament*, United Kingdom: Signal Books, p. 196

43 Scot D. Sagan, "Why Do States Build Nuclear Weapons: Three Models in Search of a Bomb," *op. cit.*, p. 68.

44 Ashok Kapur, "India's Nuclear Presence," *op. cit.*, no. 4, p. 460.

45 *Ibid.*, p. 461.

46 *Ibid.*

47 Quoted in Rajesh M. Basrur, March 2000, "Nuclear Weapons and Indian Nuclear Strategic Culture," *Journal of Peace Research*, vol. 38, no.2, p. 189.

48 Quoted in Deepa M. Ollapally, December 2001, "Mixed Motives in India's Search for Nuclear Status," *Asian Survey*, vol.41, no. 6, p. 933.

49 Praful Bidwai, May 24 - June 6, 2008, "Saying No to Nukes," *Frontline*, vol. 25, issue 11.

50 Quoted in Praful Bidwai and Achin Vanaik, *New Nukes: India, Pakistan and Global Nuclear Disarmament, op. cit.*, p. 187.

51 Weixing Hu, Summer 2000, "New Delhi's Nuclear Bomb: A systemic Analysis," *World Affairs*, vol. 163, no. 1, P. 30.

52 *Ibid*.

53 Dipanker Banarje, May 16-22, 1998, "Budha's Smile and National Security," *Economic and Political Weekly*, vol. 33, no. 20. p. 1160.

54 L. Ramdas, July 4-17, 1998, "Pokhran and its Fall Out," *Frontline*, vol. 15, no. 14.

55 Weixing Hu, Summer 2000,"New Delhi's Nuclear Bomb: A Systemic Analysis," *World Affairs*, vol. 163, no.1, pp. 30-31

56 L. Ramdas, "Pokhran and its Fall Out," *op. cit*.

57 Karsten Frey, 2006, *India's Nuclear Bomb and National Security*, New York: Routledge, pp. 80-81

58 Praful Bidwai and Achin Vanaik, *New Nukes: India, Pakistan and Global Nuclear Disarmament*, op. cit., p. 63.

59 *Ibid*., p. 63.

60 Karsten Frey, *India's Nuclear Bomb and National Security, op. cit*., p. 98.

61 M.V. Ramana and C. Rammanohar Reddy (eds.), 2003, *Prisoners of Nuclear Dream*, New Delhi: Orient Longman, pp. 44-50.

62 Praful Bidwai, "Saying No to Nukes," *op.cit*.

63 Weixing Hu, "New Delhi's Nuclear Bomb: A Systemic Analysis," *op. cit*., pp. 32-33.

64 Quoted in R. Swaminathan, 17 May 2003, "Pokhran –II: Five Years Later," Paper presented at an international session of Observer Research Foundation, Chennai.

65 Pranay Sharma, "Controversy Pokhran II: Did Our 'H' Bomb?." Available at http://www.outlookindia.com.articlesaspx?261655/ , (accessed on 30 September 2011).

66 Quoted in *Ibid*.

67 Quoted in Buddhi Kota Subba Rao, June 6-19, 1998, "The H Bomb Issue is Crucial", *Frontline*, vol. 15, no. 12.

68 Dr. B. K. Subbarao, "Our Nuclear Scam," p. 29. Available at http://www.mamushi.in/docs.com/2000/285%20vnaik/, (accessed on 5 October 2011).

69 Pranay Sharma, "Controversy Pokhran II: Did Our 'H' Bomb?," *op. cit*.

70 Buddhi Kota Subba Rao, "The H-Bomb Issue is Crucial", *op. cit*.

71 B. K. Subbarao, "Our Nuclear Scan," *op. cit.*, p .30.

72 K. Santhanam and Ashok Parthasarathi, 2009, "Pokhran-II Thermonuclear Test, a Failure," *The Hindu*, 17 September.

73 P.K. Iyengar, 2009, "Pokhran-II Revisited," *Business Daily*, 18 December.

74 Quoted in Rahul Roy Chaudhury, May 2009, "India's Nuclear Doctrine: A Critical Analysis," *Strategic Analysis*, vol. 33, no. 3, p. 405.

75 Ibid.

76 Quoted in P. R. Chari, Fall-Winter 2000, "India's Nuclear Doctrine: Confused Ambitions," The *Non-Proliferation Review*, p. 127.

77 "The Cabinet Committee on Security Reviews Operationalisation of India's Nuclear Doctrine," Press Release of 4 January, 2003.

78 Colonel Gurmeet Kanwal, "India's Nuclear Doctrine and Policy." Available at http//www.idsa.org/on-feb-1-01html, (accessed o 24 September 20011).

79 *Ibid.*

80 B. D. Jayal, "Indian Nuclear Doctrine: A Discussion." Available at http:// www. bharat.rakshak.com/monster/issue 2-3/jayal.html, (accessed on 22 September 2011).

81 Bharat Karnad, 2008, *India's Nuclear policy*, London: Praeger Security International, p. 88.

82 *Ibid.*, p. 91

83 *Ibid.*, p. 93.

84 Mohammed Badrul Alam, "India's Nuclear Doctrine: Context and Constraints," p. 12. Available at http://casi.sas.upon.edu/system/files/Heidelberg+paper+-+2002.pdf, (accessed on 22 August 2011).

85 Rajesh Basrur and Bommakanti Kartik, March 2011, "The India-China Nuclear Relationship," *Strategic Analysis*, vol. 35, no. 2, p. 189.

86 Rajesh M Basrur, 2006, *Minimum Deterrence and India's Nuclear Security*, California: Stanford University Press, pp. 27-32.

87 Gurmeet Kanwal, "India's Nuclear Doctrine and Policy," *op. cit.*

88 "How Many Nuclear Warheads Does India Need." Available at http//www.ipcs.org/article/nuclear/howmany-war-heads-does-india-need, (accessed on 20 October 2011.

89 Kanti Bajpai, "India's Nuclear Posture after Pokhran II," p.18.

Available at http://www.ceri-sciencespo.com/archieve/jan01/pokhranpdf, (accessed on 26 September2011).

90 K. S. Manjunath and Beryl Anand, February 2006, "India's Credible Minimum Deterrence," *IPCS Special Report 13*, Institute of Peace and Conflict Studies, pp. 1-4.

91 *Ibid.*

92 Quoted in Bharat Karnad, *India's Nuclear Policy, op. cit.*, p. 89.

93 V. N. Khana, 2000, *India's Nuclear Doctrine,* New Delhi: Samskriti, p. 151.

94 Namrata Goswami, July - September, 2006, "The Essence of South-Asian Nuclear Debate," *Strategic Analysis*, vol. 30, no. 3 *op. cit.*, P. 668.

95 *Ibid.*, P. 670.

96 Seema Sridhar et al, "Promoting Strategic and Missile Stability in Southern Asia," *IPCS Special Report 17*, Institute of Peace and Conflict Studies, p. 14

97 "India's Missile Programme." Available at http://www.defencepk/forums/India-defence/37481-india's-missile-programme.html, (accessed on 1 November 2011).

98 Reshmi Kazi, September 2009 "India and Nuclear Testing," *Strategic Analysis,"* vol. 33, no. 5, p. 630.

99 Thomas Mathew, Summer 2008, "Sign Of Emerging Third Leg: Strengthening Of India's Triad," *Journal of Defence studies*, vol. 2, no. 1, pp. 150-156.

100 M. V. Ramana, 2003, "India's Nuclear Command Authority," Daily Times, 9 January.

101 Suvrat Raju, January 2010, "India's Atomic Energy programme: Claims and Reality," *Aspects of India's Economy*, no. 48.

102 Ragunath Panigrahi, October 2010, "India's Civil Nuclear Diplomacy," *Civil Service Chronicle*, vol. XXI, no. 4, p. 24.

103 Anshu Bhardwaj et al, "Nuclear Power in India: The Road Ahead," *SCTEP Technical Report*, Center for Science, Technology and Policy, p. 17.

104 Richard Rhodes and Dennis Beller, January/February 2000, "The Need for Nuclear Power," *Foreign Affairs*, vol. 79, no. 1, p. 37.

105 "Development of Nuclear Energy Sector in India," *IDSA Taskforce Report No. 201*, P. 23.

106 R. G Bucher, "India's Baseline Plan for Nuclear energy Self sufficiency,"January 2009, p. 1. Available at http://

www.ipd.anl.gov/anlpub./2010/05/67057.pdf, (accessed on 23 November 2011).

107 S. K. Jain , November 2-3, 2004, "Nuclear Power In India –The Fourth Revolution," An *International Journal of Nuclear power*, vol. 18, no. 2-3, pp. 13-14.

108 R. G. Bucher, "India's Baseline Plan for Nuclear Energy Self sufficiency," *op. cit.*, pp. 2-3

109 S. K Jain, "Nuclear Power: An Alternative," p. 4. Available at http//www.npcil.mic.in/pdf/nuclear%20power%20/an%20alternatve.pdf, (accessed on 18 November2011).

110 Ian O corner, "India's Clever Nuclear Power Programme." Available at http//leadenergy.org/2011/03/indias-clever-nuclear-power-programme, (accessed on 18 November2011).

111 "Nuclear Power in India." Available at: http://www.world-nuclear.org/info/Country-Profiles/Countries-G-N/India/#.UXzs8rVTAnE, accessed on (20 November2011).

112 Anshu Bhardwaj et al , "Nuclear Power in India: The Road Ahead," *op.cit.*,pp.2-3.

113 "Nuclear energy in India: A Market Overview," November 2008, p. 2. Available at http://www.bostanalysis.com/leading_thoughts_research/Summary%20/Nuclear%20Market%20in%20India%20BA%Report.pdf, (accessed on 22 October 2011.

114 Suvrat Raju, "India's Atomic Energy programme: Claims and Reality," *op. cit*.

115 Anshu Bhardwaj et al, "Nuclear Power in India: The Road Ahead," *op. cit.*, pp. 22-23.

116 Ibid., pp. 28-30.

117 Suvrat Raju, "India's Atomic Energy programme: Claims and Reality," *op.cit*.

2

Indo-US Civil Nuclear Deal : Genesis and Development

INTRODUCTION

The cooperation between India and the United States in the field of civil nuclear energy has a very long history. Its origin can be traced right from 1950s when US offered support to India's nascent nuclear programme under "Atoms for Peace" programme. But this cooperation could not last long and suffered a setback in 1974 because of India's covert weaponisation. After decades of cold and strained relations, it was US President George W. Bush who took cudgels to transform the very essence of Indo-US bilateral relationship. He took a number of initiatives to transform the relationship between two countries but the most important and bold step was his offer of Civil Nuclear Deal to India which turned out to be an important milestone in Indo-US bilateral relationship. But in order to materialize the initiative, both the sides have to fight a tough battle. In its long journey of three years, the deal had to go through several ups and downs and had to face some tough opponents in both the countries.

This chapter traces the origin and history of Indo-US Civil Nuclear Cooperation Agreement and gives a detailed account of the process by which the initiative came to fruition. It also identifies different motivational factors behind the deal and provides a deep insight into the domestic discourse on the deal.

INDO-US NUCLEAR RELATIONS: A BRIEF OVERVIEW

"Nuclear factor" is not a new facet in India-US strategic relations but their cooperation in the field of nuclear energy has a long history. The cooperation between two sides in the field of nuclear energy started in 1950s when US extended support to India's nascent nuclear energy programme under "Atoms for Peace" programme. This programme was launched by US President Dwight D. Eisenhower to expand cooperation in the field of atomic energy. In 1956, US Atomic Energy Commission signed an agreement with India under which US agreed to supply twenty-one tons of heavy water to India for a Canadian supplied 40 MW research reactor CIRUS. In 1963, two sides signed an agreement for the construction of two Boiling Water Reactors (BWRs) at Tarapur and under a contract signed on May 17, 1966, it was also agreed upon that US will supply fuel for these twin reactors throughout their life time[1]. However, the relations between two sides started straining in mid-1960s when India refused to join Nuclear Non-Proliferation Treaty and worsened in 1974 when India carried out a Peaceful Nuclear Explosion (PNE) at Pokhran. With this PNE, US ceased all nuclear cooperation with India and put brakes on the supply of fuel to Tarapur reactors notwithstanding the fact that US was under contractual obligations to supply fuel for Tarapur reactors throughout their life time. Moreover, the 1963 agreement did not prohibit any type of PNE. India protested against this act of non-compliance as US failed to restore fuel supply and in 1980, US completely stopped the fuel supply and thus left the agreement virtually dead.[2]

The decade of 1980s witnessed no process of rapprochement and relations virtually remained cold. In this period, the two sides remained at logger heads over the nuclear issue. However, with the disintegration of U.S.S.R and end of Cold War which not only challenged the rationale for the continuation of India's policy of non-alignment but also greatly altered the strategic calculus of India, the seeds of engagement were sown as both the sides found enough reasons to redefine their relationship. India, on her part, realised that forging close relations with US which emerged as sole super power of the world after the disintegration of U.S.S.R could help it to fill the vacuum created by the disintegration of U.S.S.R[3] and US on her part realised the strategic importance of India and found India's large untapped market quite appealing. The decade of 1990s witnessed many developments towards rapprochement. In 1992, first meeting of the Indo-US Army Steering Committee was called which helped to resume strategic dialogue. After this, the meetings of the Steering Committees of Air Force and Navy followed and all this paved the way for Indo-US Military Cooperation Agreement of 1995.[4] But the process of normalisation could not lost long and ruptured in 1998 when India conducted five nuclear tests at Pokhran desert site. The tests were not less than a shock to non-proliferation priests in US and sent a clear message to them that the US policy of "non-proliferation" and "isolation" vis-a-vis India had resulted in utter failure. So, in order to bring India close to non-proliferation regime and to give a new direction to Indo-US relations, Strobe Talbott, the then US Deputy Secretary of State, started a series of talks with Indian External Affairs Minister Jaswant Singh. The two officials met around fourteen times in different parts of the world and discussed number of issues with "nuclear issue" as the main focus. Though Strobe-Jaswant spirit could not cut much ice on this issue but facilitated the visit of President Bill Clinton to India in 2000.This was the time when a US President was visiting India after a gap of twenty-two years. But Clinton's visit could not do much to repair the damaged Indo-US relations because he came to India with a "non-proliferation" agenda which greatly limited the scope to explore the potential of full cooperation between two sides. In his address to the joint session of Indian Parliament, he said:

> *"Nuclear future is not a secure future and most of the world is moving towards the elimination of nuclear weapons. That goal is not advanced if any country in any region move in the other direction."*[5]

In his speech, he questioned the rationale behind the tests and asked:

> "Was India more safe than before the tests or after the tests? Would India benefit from expanding weapons and missiles if its neighbourhood responds by doing the same? Whether India can afford to expand its nuclear and conventional forces, while meeting the requirements of development? Whether its nuclear weapons advance the large cause of humanity."[6]

During his visit, he discussed the scope of cooperation in economic field, counterterrorism and regional stability but showed no signs to break the dead lock on the nuclear issue which for decades had dogged the relations between two countries. After the visit of Clinton, John Holum, a senior adviser of Clinton on non-proliferation announced that "full potential of Indo-US relations couldn't be realised until India joined NPT." [7]

A paradigm shift in Indo-US relations came in 2000 when George W. Bush came to power in U.S. After assuming the office of the President, he reversed the US policy of "isolation" vis-à-vis India and put emphasis on the policy of "engagement" because it was a long-term objective of George Bush to engage world's largest democracy, often telling his advisers that "You can't ignore one billion strong democracy. We have to find ways of engaging with it."[8] Even before his presidency, it became evident that Bush is determined to transform Indo-US bilateral relationship as Condoleezza Rice, who served as National Security Adviser during Bush's first term indicated during 2000 presidential campaign that "a future Bush Administration would take a new approach to India." In the same year, she wrote in *Foreign Affairs* journal that:

> *"India is not a great power yet, but it has the potential to emerge as one... India is an element in China's calculation and it should be in America's too."*[9]

Robert Blackwell, who served as US Ambassador to India from September 2001 to July 2003 while speaking at the Indo-American Chamber of Commerce on September 6, 2001 also said that Bush administration was determined to "transform fundamentally the very essence of bilateral relationship."[10] Within these developments erupted September 11, 2001 crises which proved a blessing in disguise for India. 9/11 crises made Bush administration to shift its focus from non-proliferation to terrorism which was a common enemy to both the countries. In its fight against terrorism, India extended full support and cooperation to Bush administration. It supported US in operation "Enduring Freedom" by escorting US warships through Malacca Strait. India also supported US in Iraq war, [11] granted permission to FBI to set up its office in New Delhi and also offered ports and air port facilities to US.

In order to take the process of cooperation forward and to widen its scope, President Bush and Indian Prime Minister Atal Bihari Vajpayee announced Next Steps in Strategic Partnership (NSSP) in January 2004 which turned out to be a milestone in emerging Indo-US relations. Under NSSP, both the sides agreed to take certain reciprocal measures to facilitate cooperation in technology transfers. US agreed to ease restrictions on the transfer of technology to India in three crucial areas, civil nuclear technology, civilian space and dual-use technology goods. India on its part agreed to tighten the control on the export of sensitive technology and not to divert technology received from US towards military purposes.[12] First phase of NSSP was concluded in September 2004 when Indian Foreign Secretary Shyam Saran and US Under Secretary of State Kenneth Juster sealed an agreement under which Indian Space Research Organisation (ISRO) was removed from "Entity List" and licence requirements for exporting low level dual use items to Indian research organisations were also eased.[13] After this land mark initiative, several other positive developments took place like in October 2004, officials of the two sides held two day talks in Delhi to improve cooperation in high technology defence trade, space programmes and nuclear sector. This was followed by the visit of Head of the US Nuclear Regulatory Commission in February 2005 to India[14] which greatly helped to transform Indo-US relations, however, most significant development was the visit of US Secretary of

State Condoleezza Rice to India. Rice visited India in March 2005 and expressed willingness of Bush administration to cooperate with India in the field of civil nuclear energy. Indians were highly surprised by this offer and lost no time in seizing the opportunity and immediately after her visit, two sides started negotiations to chalk out contours for civil nuclear cooperation. After some high profile talks and delegation level meetings, Indian Prime Minister Dr. Manmohan Singh and US President George Bush issued a joint statement on 18 July, 2005 which set the framework for Indo-US Civil Nuclear Cooperation Agreement. With this initiative, Bush not only reversed decades old US policy on non-proliferation but also removed a biggest obstacle, which for years had clouded the relations between two countries. In the joint statement, President Bush stated that "as a responsible state with advanced nuclear technology, India should acquire the same benefits and advantages as other such states."[15] President Bush also assured Prime Minister Dr. Manmohan Singh that "he will work to achieve full civil nuclear energy cooperation with India as it realizes its goals of energy security."[16] In order to realise the potential of full civil nuclear engagement, Bush also announced that:

- He will seek an agreement from US Congress to adjust US laws to enable full civil nuclear cooperation with India.
- He will work with friends and allies to adjust international regimes to realise the goal of full civil nuclear cooperation.
- US will work towards the inclusion of India in ITER project and Generation IV International Forum.[17]

On his part, Indian Prime Minister announced that "India would reciprocally agree that it would be ready to assume the same responsibilities and practices and acquire the same benefits and advantages as other leading countries with advanced nuclear technology, such as United States."[18] These responsibilities and practices would include:

- Identifying and separating India's civil and military nuclear facilities.
- Placing the civil nuclear facilities under the safeguards of International Atomic Energy Agency (IAEA)
- Negotiating an Additional Protocol with International Atomic Energy Agency.
- Continuing its unilateral voluntary moratorium on nuclear testing.
- Working with US towards the conclusion of Fissile Material Cut off Treaty.
- Exercising a strict control on the transfer of sensitive nuclear technologies.[19]

OPERATIONALISATION OF THE NUCLEAR DEAL

Indo-US Civil Nuclear Deal was a complex deal by nature as it had to go through several complex stages to see the light of the day which included:

- Creating a wall of separation between India's civil and military nuclear facilities (Separation Plan).
- Amendment of US domestic law.
- Negotiating a bilateral 123 Agreement.
- Negotiating a Safeguards Agreement with IAEA.
- Adjustment of the guide lines of Nuclear Suppliers Group (NSG).

As a first step towards the operationalisation of the nuclear deal, India immediately after the joint statement started negotiations with US interlocutors to chalk out a "separation plan", splitting India's nuclear facilities into military and civilian ones. But it was not altogether an easy task as India had to design a comprehensive plan satisfying the interests of both the parties. It had to be a balanced plan: not affecting India's strategic nuclear programme on the one hand and not hurting US from non-proliferation perspective on the other hand. US President urged India to "produce a credible, transparent and defensible plan to separate its civilian and military nuclear programmes under the same international safeguards that govern nuclear power programmes in other countries."[20] During negotiations, Indian breeder programme remained a contentious issue. While US was insisting India to place its maximum number of reactors including breeder reactors under safeguards but Indian position was to place minimum number of reactors under safeguards and keep its breeder programme outside the purview of International Atomic Energy Agency. After much bargaining, two sides finally reached an agreement which was concluded during the visit of President George W. Bush to India in March 2006. Under the agreement India agreed to place fourteen out of twenty-two reactors in operation/construction in the civilian list. These reactors will be placed under the safeguards of IAEA in a phased manner. But it was also agreed upon that the safeguards will not apply to India's breeder nuclear programme.[21]

After the announcement of "separation plan", Bush administration passed "Henry J. Hyde United States-India Peaceful Atomic Energy Cooperation Act of 2006" also called Hyde Act on 18 December, 2006. The Act provides legal basis for nuclear cooperation between India and United States. But the passage of the passage of the Act created uproar among the India scientific and political communities because of certain controversial provisions. The Act limits the "reprocessing rights" of India on the spent fuel and calls for cessation of nuclear cooperation with India if India conducts a nuclear test. Further, the

Act requires India to work with US in its non-proliferation efforts particularly in containing Iran's nuclear programme. These provisions are contrary to the assurances given by Indian Prime Minister to Parliament on civil nuclear cooperation with US on different occasions. The passage of the Act generated a storm of controversy in India and even haunted the fate of Civil Nuclear Cooperation Agreement. All this heightened the concerns of Indian Prime Minister as he was reported as saying on Zee News: "There are issues. There are concerns we have to discuss with US administration. Some of these things will be discussed in 123 (bilateral) agreement on which negotiations will be shortly held."[22] Expressing a deep sense dissatisfaction over the provisions of the Act Shyam Saran said that, "Negotiations to conclude a bilateral agreement on civil nuclear cooperation with US will be very difficult... A number of key Indian concerns will have to be met or the agreement ... could fall through."[23]

When negotiations on 123 Agreement started, Indian negotiators had two choices; either to strike such an Agreement which could address their concerns and which they could sell in India or to run away. However, Indian negotiators choose the former and after some nine rounds of hectic and complex negotiations, two sides struck a bilateral 123 Agreement stipulating the terms and conditions under which nuclear trade between the two sides would take place. The Agreement according to Indian negotiators addressed their three main concerns, viz. "reprocessing of spent fuel generated from imported reactors", "fuel supply assurances", and "non-interference in India's strategic nuclear programme". The Agreement was settled on 20 July 2007 and text of the Agreement was released on 3 August 2007.[24] Nicholas Burns, then Undersecretary of State for Political Affairs and a chief negotiator of the Agreement called the Agreement "the symbolic centrepiece of a growing global partnership between our two countries [India and the United States]."[25] In a joint statement Indian External Affairs Minister Pranab Mukherjee and US Secretary of State Condoleezza Rice termed the Agreement as a "Historic milestone in [the] strategic partnership between two countries."[26] While addressing a joint press conference after finalising the Agreement, National Security Adviser, M. K. Narayanan and Chairman of the Atomic Energy Commission, Anil Kakodkar said that the Agreement "was fully consistent with the commitments undertaken by both the countries in the Joint Statement of July 18, 2005 and March 2, 2006."[27] In his remarks Anil Kakodkar described the Agreement "satisfactory" and said that the Agreement gives India advance consent rights to reprocess spent fuel and recover material from it in a safeguarded national facility. On fuel supply assurances, Kakodkar said that, "the fuel supply assurances built into March 2006 Statement and in this final draft are entirely consistent" and the Agreement does not interfere with country's strategic nuclear programme.[28]

After finalising the 123 Agreement, India negotiated a Safeguards Agreement with International Atomic Energy Agency in August 2008 and in the same year, Nuclear Suppliers Group adjusted its guidelines for India to engage it in nuclear commerce with other counties of the world. Thereafter, the Agreement was sent to US Congress for its approval where it faced stiff opposition. However, after much debate and discussion House of Representatives endorsed it on 28 September, 2008 with 298-117 votes. The Senate approved it on 1, October, 2008 with 86-13 margin. US President signed the legislation on the nuclear deal into law on 8 October, 2008, now known as, "United States-India Nuclear Cooperation Approval and Non-Proliferation Enhancement

Act". At the signing ceremony, Bush stated that the legislation does not make any change in basic parameters of the 123 Agreement that he had sent to the Congress. While allaying the Indian concerns on "fuel supply assurances" Bush said that:

> *"the legislation does not change the fuel supply commitments the US government has made to the government of India as recorded in the 123 Agreement."*[29]

It may be noted that such concerns were raised when President Bush wrote a letter to US Congress in September 2008, stating that the fuel supple assurances as laid down in the bilateral 123 Agreement are merely 'political' rather than legally binding commitments.[30] In his statement, Bush also underlined that the Agreement grants India 'advance consent to reprocessing' "which be brought into effect upon the conclusion of arrangements and procedures for a dedicated reprocessing facility under safeguards."[31]

The Agreement was formally inked by Indian External Affairs Minister, Pranab Mukherjee and his counterpart US Secretary of State, Condoleezza Rice on 10 October, 2008. The signing took place in the Benjamin Franklin Room of the US State Department. At the signing ceremony, Indian External Affairs Minister, Pranab Mukherjee said that the Agreement represents a sign of the transformed relationship between India and the United States. Stressing the binding nature of the Agreement, he said that:

> *"The Agreement has been passed by the U.S. Congress without any amendments. Its provisions are now binding on both the sides once the Agreement enters into force."*[32]

In her opening remarks, US Secretary of State

Condoleezza Rice praised the Indian Prime Minister for risking his political future for the sake of the Agreement. She called the Agreement "a diplomatic triumph" for both the countries and said that the Agreement unlocked "a new and a far broader world of potential for our strategic partnership in the 21st century, not just in nuclear cooperation but in every area of national endeavour."[33]

MAIN FEATURES OF THE AGREEMENT

- The Indo-US Civil Nuclear Cooperation Agreement provides for civil nuclear cooperation between India and the United States in the areas covering nuclear reactors and other aspects of nuclear fuel cycle including technology transfer on industrial or commercial scale. Besides this, the Agreement also provides for cooperation in biological research, medicine, agriculture and industry, environment and climate change (Art. 2).[34]
- Agreement gives India "right to reprocess" US origin spent fuel. For this purpose, India will establish a reprocessing facility where reprocessing of fuel will take place under the monitoring of International Atomic Energy Agency. India and US will mutually agree on arrangements and procedures under which reprocessing will take place in the safeguarded facility (Art. 6(iii).[35]
- The Agreement gives India "fuel supply assurances" for reactors placed under IAEA safeguards. United States will support India to build up a "strategic reserve of fuel" to guard against any disruption of fuel. Despite this if any disruption of fuel takes place, India and US will jointly convene a group of suppliers to restore fuel supply to India. Despite all these arrangements if disruption in fuel supply occurs India has a right to take "corrective measures" (Art. 5).[36]

- The Agreement grants 'right of return' to both the parties. In case of termination of the Agreement, Art.14.4 says that the either party shall have the right to seek "return of any nuclear material or components transferred under this Agreement and any special fissionable material produced through their use." But recognising that the invoking of this right could have serious implications on the relations of two parties, Agreement provides for several layers of insulation including holding consultations giving "special consideration to the importance of uninterrupted operation of nuclear reactors of the concerned party." Besides this "fair market value" compensation must be paid by the party invoking "right of return" and agreement must be reached on "methods and arrangements" for return of material and equipments (Art 14.6). [37]
- The Agreement does not take away India's "right to test" a nuclear device in future. In fact, on the issue of "right to test" the Agreement is silent, but it is assumed that India can conduct a nuclear test, if her security interests are jeopardised. But once India conducts a nuclear test, US has right to serve notice of termination of the Agreement under Article 14.1. But Article 14.2 of the Agreement says that before terminating the Agreement two sides would hold consultations on the circumstances which prompted India to take such a decision, including a change in her security environment or whether the test was a response to a similar action by states affecting India's national security interests. If the decision was a response to latter action, then consultations would end and notice of termination would be withdrawn, if it was an unprovoked test India still can claim that it was her changed security environment which

prompted her to take such a step[38] but in such a scenario US will have final word on the termination of the Agreement.

- The Agreement does not interfere with India's strategic nuclear programme. The Agreement provides for the application of safeguards only on imported material and reactors and only on those indigenous reactors which India declares as civilian.
- The Agreement ends India's decades old nuclear isolation and segregation. It opens the doors of cooperation for India in the civil nuclear energy field and provides India access to nuclear fuel, equipment and technology.
- The Agreement will remain in force for a period of 40 years, but it can be further extended for 10 years. It can be terminated by either party even before its expiration by giving a one year notice. The party invoking the right of termination will have to provide the reasons for termination (Article14.1).

OBJECTIVES BEHIND THE DEAL

The Indo-US Civil Nuclear Cooperation Agreement is not only about cooperation between India and the United States in the field of civil nuclear energy but it is more than that. It has a strategic intent and is aimed at to carry out certain strategic objectives as was outlined by US Secretary of State Condoleezza Rice at the Agreement signing ceremony:

> *"Let no one assume that our work is finalised. Indeed what is most valuable about this agreement is that it unlocks a new and a far broader world of potential for our strategic partnership in 21st century, not just on nuclear energy but on every area of national endeavour."*[39]

Both the sides look at the Agreement as a vital instrument to achieve certain strategic goals. The US considers it important to carry out four broad strategic goals.

BALANCING CHINA

Since the end of the Cold War, it has been US policy to remain engaged in every part of globe and not to allow any particular power to dominate any particular region. This was confirmed by US Defense Planning Guidance (DPG) document published in 1992. It notes that: "to prevent the re-emergence of new rival" is the primary objective of the country and not to allow any hostile power to dominate a region and "generate global power".[40] The rise of China as a regional power after the disintegration of USSR created serious apprehensions for US as it started viewing China as a serious security threat. Bush administration directly dubbed China a "future military adversary" and an important "strategic issue" facing US.[41] The US Defense Department in its 3rd quadrennial review released in February, 2006 described China as "potential rival".[42]The rise of China as a great economic and military power in the region has not only created an imbalance in the region but has also greatly limited the chances of US engagement in the region, thus posing a direct challenge to US. To quote Kapil Sibal, Union Minister for Science and Technology: "[I]f the US faces a challenge in the 21st Century, it will not be from India; (but) somebody from its neighbourhood."[43] So, in order to face this challenge, US has offered nuclear deal to India by which US not only wants to balance the rising power of China but also wants to make its presence felt in the region from which it has been excluded from decades. According to Ferguson, Science and Technology fellow at Council on Foreign Relations, "The United States is trying to cement its relationship with the largest democracy in order to counter balance China."[44] Billl Emmott, Editor in Chief of *The Economist* also writes that, "the basic reason to make India an exception and to bring it closer to US is the desire to balance rising power of China in Asia."[45]

ISOLATING IRAN

Second Important US objective is to isolate Iran because of her controversial nuclear programme. The US domestic legislation Hyde Act calls up on the US administration to secure India's full support in its "Iran isolation efforts". Section 103 (b) (4) of the Act states that with respect to South Asia it shall be the policy of United States to:

> *"Secure India's full and active participation in United states efforts to dissuade, isolate, and, if necessary, sanction and contain Iran for its efforts to acquire weapons of mass destruction, including a nuclear weapons capability and the capability to enrich uranium or reprocess nuclear fuel, and the means to deliver weapons of mass destruction."*[46]

However, this US policy vis-a-vis Iran raises very important question: Why US has chosen India as a junior partner in its "Iran Isolation efforts"? Possibly this is because of close trade and cultural links between Iran and India and India's continued assistance to Iran's nuclear programme. In order to further strengthen their cooperation, two sides launched a "strategic partnership" in 2003 and India also reiterated its commitment that it would continue to assist Iran's nuclear programme. This created serious apprehensions among US officials. During a hearing on Indo-US relations in September 2005, a number of members of House International Relations

Committee expressed their serious concerns on the close India-Iran relations. But what irritated US most was the desire of India to participate in multi-billion dollar IPI (Iran-Pakistan-India) gas pipeline project.[47] Realising that the project would pump billions of dollars into Iran and would undermine the US efforts to isolate Iran, US offered nuclear deal to India to prevent her participation in the project. During a hearing on the deal US, Secretary of State Condoleezza Rice said that "the civil nuclear cooperation with India would move New Delhi away from seeking oil and natural gas from Tehran."[48] But the real objective was not to reduce India's reliance on Iranian oil and gas but to put an end to their emerging strategic cooperation and to secure India's support against Iran at IAEA to refer her to UN Security Council because of her controversial nuclear programme. Then US Ambassador to India, David Milford clearly declared to Indian officials that the civil nuclear deal would die if India did not support US backed resolution at IAEA against Iran: "I think Congress will stop considering the matter. I think the initiative [Civil- Nuclear Cooperation Agreement] will die in the Congress..."[49]

TO BRING INDIA CLOSE TO NON-PROLIFERATION REGIME

Third important US objective behind the deal is to bring India close to NPT regime. It has always been a major foreign policy challenge for US to bring India within the ambit of NPT which India considers highly discriminatory. Realising that past US policy of isolation vis-à-vis India not only kept India outside the NPT regime but provoked India to accelerate her weapons programme, the US policy makers adopted a policy of engagement to bring within the ambit of NPT and for this purpose nuclear deal is being considered an important instrument. To quote US Security Advisor Hedley:

> *"this [civil nuclear deal] is actually a way to bring India within non-proliferation framework. It will be adopting the kind of non-proliferation protection that we do. It is to bring India on the same page as we, so they can be global partners in the proliferation issue as well... This is actually the vehicle to bring India on to our page in terms of proliferation policy."*[50]

Under the Agreement India has agreed to divide its nuclear programme in to military and civilian ones and has also committed to place its fourteen reactors under the IAEA safeguards. This will make India to open 65% of its reactors for IAEA inspections so that nuclear material and technology is not diverted. This figure could rise to 90% in future. Prior to the Agreement, only 19% of India's nuclear reactors were open for inspections. Besides this, India has also committed to continue its unilateral moratorium on nuclear testing, negotiate a multilateral Fissile Material Cut off Treaty and not to transfer sensitive nuclear technology to other countries. All these steps bring India close to NPT regime. According to Dr. Mohamed ElBaradei, former Director of IAEA:

> *"[T]he nuclear deal would bring India closer as an important Partner in the non-proliferation and which would be a milestone, timely for ongoing efforts to consolidate the non-proliferation regime, combat terrorism and strengthen nuclear safety."*[51]

TO SCORE ECONOMIC GAINS

The nuclear deal has a great potential to boost economic fortunes of United States. The deal will open Indian nuclear market for American firms and will generate employment opportunities for Americans. As India plans to build at least twenty-four nuclear reactors in next fifteen years with foreign assistance and even a small share of this project could earn billions of dollars for American nuclear vendors and would generate employment opportunities for Americans. This was confirmed by Jeffery T. Bergner, Assistant Secretary of Legislative Affairs before the House Committee on Foreign affairs:

> *"Of the 15 or so reactors India plans to import, even if the US vendors can win two reactors it could add 3000-5000 direct Jobs and 10,000-15,000 indirect job in America for Americans. Furthermore, at least 15 US companies including General Electric and Westinghouse will be involved in the process."*[52]

The nuclear deal is also projected as a boon for American nuclear sector as it will rejuvenate American nuclear industry which is lying dormant since Three Mile Island accident and is losing markets in the world. The deal will open new avenues for American nuclear Industry and will make it globally more competitive. This was admitted by Condoleezza rice in these words:

> *"At its core, our initiative with India is not simply government to government effort. It was crafted with private sector firmly in mind. Because it will fully open the door to civil nuclear trade and cooperation, this initiative is good for American business."*[53]

Not only the nuclear sector but other sectors like aerospace and defense are also likely to reap huge benefits. The order for purchasing commercial air planes worth $ 8 billion from Boeing of Seattle has been placed before it. Likewise, American defense industry by selling military equipments and hardware to India could gain billions of dollars. William Cohen, former US Defense Secretary, Claimed that "the US defense companies could gain purchases from India worth $ 30 billion in the period 2007-2012."[54]

Like the US, India also looks at the deal as an important instrument to carry out certain important goals these include:

END OF NUCLEAR ISOLATION

One of the motivational factors behind India's engagement with US in civil nuclear energy field was to get rid of decades old nuclear isolation. Right after 1974, India was subjected to severe international sanctions because of her nuclear test which left India isolated in the international environment. Export of nuclear technology and material was highly restricted to India. Such was the state of affairs that India was not allowed to purchase a super computer for weather forecasting because of its dual use. Indian nuclear scientists were denied visas by US and were not allowed to visit nuclear laboratories in US. All these restrictions greatly hindered the development of India's power programme. In mid-1990s, its power reactors had some of the world's lowest capacity factor, telling the story of country's nuclear isolation. Nuclear deal seeks to put an end to this nuclear segregation and technology denial regimes. The Agreement gives India access to nuclear material, equipment and dual use technology. India can now engage in civil nuclear trade not only with US but also with other countries of the world. The cooperation in the nuclear field is not only considered important for Indian nuclear secretor but other sectors like space and defense are also likely to get huge benefits. Indian space industry now can cooperate with US companies for developing and marketing US commercial satellites. In the same way, Indian denfense industry can get latest military equipments and hardware from US.

TO MEET GROWING ENERGY DEMANDS

India is one of the fast growing economies of the world with burgeoning energy requirements. In the period between 1980-2001, India witnessed 208% increase in the demand of energy, in contrast to this, China only saw 130% increase in the same period.[55] Its energy requirements are expected to increase by four times in next twenty-five years. Currently, India is fifth largest consumer of energy in the world and is expected to assume third rank in the middle of century. So, in order to meet its future demands of energy, India has to look for a robust source of energy which can quench its energy demand for a longer period of time. Currently, Coal, oil and hydro power are three main contributors to India's energy mix. But for longer time perspective these sources are not considered a feasible option. India is world's third largest producer of coal but most of the Indian coal is inefficient with high ash content. That is why India has to import a huge quantity of coal every year. In 2010-2011, it imported 82 billion tonnes of coal and this figure is likely to increase in future. Experts believe that by 2032, India will import coal worth $ 230-250 billion.[56] Besides this, coal has a very harmful effect on our environment as it generates a huge amount of greenhouse gases. India's emission of greenhouse gases increased by 61% between 1990-2001 and as per the estimates of Department of Atomic Energy, India's carbon emissions will increase by 3% annually between 2001 and 2025.[57] Hydropower is a clean source of energy and India has a vast hydropower potential but problem with the exploitation of this source is that the hydropower projects cause damage to environment and displacement of large sections of population. Oil is a significant contributor to India's energy basket. It accounts for nearly 33% of India's energy consumption but India has very meagre reserves of oil and about 70% of its requirements are met through imports. India wants to free itself from heavy reliance on oil imports because of instability of oil prices in international market. So, keeping in view the limitations associated with the use of these sources, India policy makers have chosen nuclear route to meet the future demand of energy. Among all the possible sources of energy, nuclear energy has got the highest potential. Besides this, it is also considered a clean source of energy as it does not produce greenhouse gases. As per the estimates of Nuclear Energy Institute, the nuclear power plants in US in the year 2004 "prevented 3.43 million tonnes of sulphur dioxide, 1.11 million tonnes of nitrogen oxide, 696.6 million metric tonnes of carbon dioxide from entering earth's atmosphere."[58] Votaries of nuclear energy believe that nuclear power in India can also bring same benefits and it has been estimated that 10 fold increase in India's nuclear power generating capacity would decrease India's carbon dioxide emission by 130 million tonnes annually.[59] So, the importance of civil nuclear deal is that it will not only help Indian to meet her growing energy needs by providing her access to latest nuclear technology and fuel but it will also help to reduce its heavy dependence on coal and oil which in turn can help to protect environment by reducing the emission of greenhouse gases and it is for this reason Condoleezza Rice called it "one of the greenest parts of India's new Green Revolution."[60]

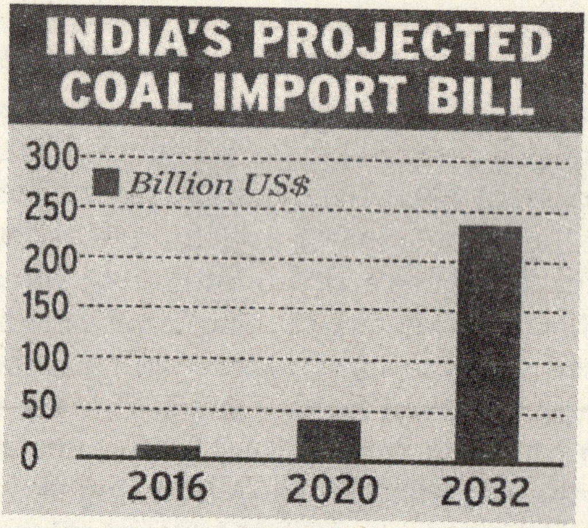

Fig. 2.1

Source: The Hindu, November 6, 2011.

FUEL FOR REACTORS

Among the developing countries, India probably has largest and unique nuclear programme, with 20 operational reactors but shortage of fuel has been a main constraint for its nuclear power programme. India has very limited reserves of uranium (61,000 tonnes) which is used as fuel in Pressurised Heavy Water Reactors and access to foreign fuel was restricted to it by technology denial regimes in the wake of her 1974 nuclear test. There has been a big gap between demand and supply, for instance in 2006-2007 the demand for uranium was 475 tonnes per year but production was only 300 tonnes.[61] The shortage of fuel has been a main factor responsible for low capacity factor of its nuclear reactors. In 2000-2003, its Pressurised Heavy Water Reactors were operating at 90% capacity factor. But it dropped to 81% in 2003-2004 and worsened in 2004-2005 when it reached to 76%.[62] The nuclear deal has opened the doors for India to access fuel for its power reactors not only from US but from other suppliers as well and in this way a biggest impediment of its nuclear programme stands removed. After nuclear deal was signed, an Indian official told BBC that: "If this agreement had not come through we might have... closed down our nuclear reactors and by extension our nuclear programme" [63] because of shortage of fuel.

INDIA'S SEPARATION PLAN

Separation of Indian nuclear facilities into "civilian" and "military" ones stems from July 18 joint statement in which Prime Minister Dr. Manmohan Singh had made a commitment that India will divide her nuclear facilities into "civil" and "military" ones and will place former under IAEA safeguards. In order to chalk out a comprehensive "separation plan" India constituted two subcommittees involving key functionaries from DAE, Ministry of External Affairs, the Armed forces and Prime Minister's Office. When negotiations started, US administration made it clear to the Indian negotiators that "separation plan" should be "credible" and 'defensible" from non-proliferation stand point[64] as international legal standards and US domestic laws make it obligatory upon US that civil nuclear cooperation with a non-nuclear weapon state must not result in any sort of nuclear proliferation. On November 2, 2005 Robert Joseph, Under Secretary of State told members of Senate Foreign Relations Committee that India's "separation plan" "must be credible, transparent, meaningful and defensible from non-proliferation stand point." He further added that India must bring maximum number of facilities under safeguards and safeguards should be applied in perpetuity.[65] In December 2005 when an Indian delegation visited US, Richard Lugar, Chairman of US Senate Committee on Foreign Relations made it clear to Indian delegates that Indian "separation plan" must meet three requirements: It should be transparent, it should not assist India's weapons programme and should ensure that all the nuclear material exported to India is placed under the monitoring of IAEA.[66] But India had its own compulsions. It had to design such a plan satisfying two important requirements — it must not hurt India's national security interests and must not cripple her deterrent capability— as Prime Minister Manmohan Singh had assured Indian Parliament that:

> *"We will offer to place under safeguards only those facilities that can be identified as civilian without damaging our deterrent potential or restricting our R and D effort or in any way comprising our autonomy of developing our three stage nuclear programme."*[67]

This difference in perspectives made it difficult for negotiators to chalk out a comprehensive plan accommodating the concerns of both the parties. As the negotiations on the "separation plan" were underway, pressure started mounting on both the sides. In US, non-proliferation lobby urged Bush administration that Indian "separation plan" must address four concerns:

I. It must put a cap on India's fissile material production.

II. India must place CIRUS reactor and the plutonium recovered from it under the safeguards.

III. India must place all the thermal power reactors and breeder reactors under safeguards along with other nuclear facilities including heavy water production plants, reprocessing facilities, fuel fabrication and enrichment facilities.

IV. India must adopt permanent safeguards on its civilian facilities so that it could not remove them from safeguards in future.[68]

In India, the nuclear scientific establishment urged the government that:

- Indigenously developed nuclear facilities must be kept outside the purview of IAEA safeguards.
- Safeguards should not be permanent in nature.
- Indian breeder programme should be kept off the safeguards as it has been developed with indigenous technology and is still in technology development phase.[69]

During the negotiations "nature of the safeguards" and Indian breeder programme remained two contentious issues which could only be settled during the visit of President Bush to India in March 20006. After much bargaining, the negotiators were able to settle down an agreement. In the final agreement, it was agreed that safeguards will be permanent in nature but will apply to only those facilities declared by India as civilian. It was also agreed that safeguards will not apply to Indian breeder programme. President Bush termed the plan as "a good faith gesture" by Indian government that he can take to US Congress.[70]

The "separation plan" thus agreed was unveiled in Indian Parliament by Prime Minister Dr. Manmohan Singh on 7 March, 2006. While highlighting the main features of the plan Dr. Manmohan Singh said that:

- Under the "separation plan", India will place fourteen thermal power nuclear reactors out of twenty-two reactors in operation/construction under safeguards. The process will be completed by the end of 2014.
- Under the plan India has agreed to keep its nuclear breeder programme off the safeguards which is still in R and D phase.
- India will permanently shut down CIRUS reactor and will bring fuel core of Apsara reactor under safeguards.
- "Separation plan" does not cap India's strategic

nuclear programme. It ensures adequacy of fissile material and other inputs to meet the current and future requirements of India's nuclear programme.

- "Separation plan" guarantees reliable supply of fuel for safeguarded nuclear reactors.
- "Separation plan" protects India's credible minimum deterrence. In the words of Dr. Manmohan Singh:

  "No constraint has been placed on our right to construct new facilities for strategic programme. The integrity of our nuclear doctrine and our ability to sustain a minimum credible nuclear deterrent is adequately protected."[71]

The "separation plan" agreed in March 2006 was tabled in Indian parliament in its final form on 11 May, 2006. The separation of Indian nuclear facilities was guided by certain principles as spelt out in the document entitled "Implementation of the India-United States Joint Statement of July 18, 2005: India's Separation Plan". These principles are:

- India will open only those facilities for international inspections which after safeguards will no longer be engaged in the activities of strategic importance.
- The criteria to offer a facility for safeguards would be purely determined by national security considerations.
- A facility located in the "hub of strategic significance" will be kept off the safeguards, despite the fact that it may not be involved in the activities of strategic importance.
- Therefore, a civilian facility would be determined as one not relevant for India's strategic nuclear programme.[72]

On the basis of above mentioned principles, India proposed the following:

1. Thermal power reactors : India will identify and offer fourteen thermal power reactors for safeguards between 2006 and 2014. These reactors will include six reactors which are already under safeguards. This will open sixty-five per cent of total installed capacity for international inspections form current nineteen per cent.

2. Fast breeder reactors : India will not place Prototype Fast Breeder Reactor (PFBR) and the Fast Breeder Test Reactor (FBTR) under safeguards as the breeder programme is still in R and D phase.

3. Future reactors : India will place its future civilian thermal and breeder reactors under the safeguards but determination of such reactors as civilian will be an Indian decision.

4. Research reactors : India has decided to permanently shut down CIRUS reactor in 2010 and will move fuel core of Apsara reactor purchased from France out of BARC and will place it under safeguards in 2010.

5. Upstream facilities : in the nuclear fuel complex of Hyderabad following facilities will be placed under safeguards by 2008:

- Uranium Oxide Plant (Block A).
- Ceramic Fuel Fabrication Plant(Palletizing) (Block A)
- Ceramic Fuel Fabrication Plant (Assembly) (Block A)
- Enriched Fuel Fabrication Plant.
- Gadolinia Facility.

Besides these facilities, Heavy Water Production plants at Thal, Tuticorin and Hazira will also be placed under safeguards between 20006-2009, but these are not relevant for safeguards.

6. Downstream facilities : India will open Tarapur

Power Reactor Fuel Reprocessing Plant for safeguards after 2010 in a "campaign mode'. Tarapur and Rajasthan "Away From Reactors" spent fuel storage pools would be placed under safeguards between 2006-2009.

7. Research facilities : India will designate following research facilities as civilian, however these are not relevant for safeguards.

- Tata Institute of Fundamental Research
- Variable Energy Cyclotron Centre.
- Saha Institute of Nuclear Physics.
- Institute for Plasma Research
- Institute of Mathematics Science.
- Institute of Physics
- Tata Memorial Centre.
- Board of Radiation and Isotope Technology.
- Harish Chandra Research institute.

The "separation plan" guarantees reliable fuel supplies to India for the reactors designated as civilian and in return India will place its civilian reactors under the permanent safeguards of International Atomic Energy Agency.[73]

Table 2.1. Reactors designated as civilian and offered for safeguards under the separation plan

S. No.	Facility	Unit	Year offered for safeguards
1.	Tarapur atomic Power Station(TAPS	1	2006
2.	Tarapur atomic Power Station(TAPS)	2	2006
3	Rajasthan Atomic Power Station (RAPS)	1	2006
4	Rajasthan Atomic Power Station (RAPS)	2	2006
5	Kundankulam Nuclear Power Project (KKNPP)	1	2006
6	Kundankulam Nuclear Power Project (KKNPP)	2	2006
7	Rajasthan Atomic Power Station (RAPS)	5	2007

8	Rajasthan Atomic Power Station (RAPS)	6	2008
9	Rajasthan Atomic Power Station (RAPS)	3	2010
10	Rajasthan Atomic Power Station (RAPS	4	2010
11	Kakrapar Atomic Power Station(KAPS)	1	2012
12	Kakrapar Atomic Power Station(KAPS)	2	2012
13	Narora Atomic Power Station (NAPS)	1	2014
14	Narora Atomic Power Station (NAPS)	2	2014

(***Source:*** http//www.dae.gov.in/press/sepplan.pdf.)

Note: TAPS 1 and 2 and RAPS 1 and 2 are already under safeguards and Kundunkulam1 and 2 which are under construction are committed to be placed under safeguards.

Table 2.2. Unsafeguarded reactors

S. No.	Facility	Date of completion
1	Madras Atomic Power Station (MAPS)—1	1984
2	Madras Atomic Power Station (MAPS)—2	1986
3	Tarapur atomic Power Station(TAPS)—3	2007
4	Tarapur atomic Power Station(TAPS)—4	2006
5	Kaiga Atomic Power Station—1	2000
6	Kaiga Atomic Power Station—2	2000
7	Kaiga Atomic Power Station—3	2007
8	Kaiga Atomic Power Station¯4	2007

(***Source:*** Col. Ved Prakash, "Indo-US Civil Nuclear Deal", in Rohul Bhonsle et al (eds.), *Indo-US Civil Nuclear Deal*, Vol. II, p. 87)

AN ASSESSMENT OF THE SEPARATION PLAN

A casual reading of the "separation plan" gives the impression that the plan is balanced one and accommodates the concerns of both the parties but a deeper insight makes it clear that it favours India in a number of areas and falls short of US expectations as Undersecretary of State, Nicholas Burns told reporters on 2, March 2006, after the formulation of the plan that "It is not a perfect deal in the sense that we have not captured 100 percent of India's nuclear programme..."[74]

In the first place, plan does little to cap India's fissile material production. The plan captures only 65 per cent of the India's reactors and the remaining 35 per cent of reactors which stand off the safeguards can generate plutonium sufficient to produce 25-40 fission bombs annually. These unsafeguarded reactors have a capacity of about 2,400 MWs and each MW roughly produces 0.5 kg of plutonium and it requires about 3 to 5 kg of plutonium to develop a nuclear bomb.[75] Moreover, Dhruva reactor which is off the safeguards could produce plutonium sufficient to develop eight bombs annually and 15 MW Fast Breeder Test Reactor could generate plutonium for four bombs annually[76]. The 500 MW Prototype Fast Breeder Reactor which is still under construction could produce plutonium sufficient for dozens of warheads annually.

Secondly, "separation plan" does not cover India's entire nuclear fuel cycle of which thermal power reactors are a small component. It does not capture enrichment and reprocessing facilities which are most sensitive components of nuclear fuel cycle. Out of the four reprocessing facilities, only Tarapur Power Reactor Fuel Reprocessing Plant has been offered for safeguards which India had already been placing under safeguards in "campaign" or "temporary" mode when safeguarded fuel form safeguarded reactors was to be reprocessed in the facility.[77]

Thirdly, the nine reach facilities which have been declared as civilian are irrelevant for safeguards. A facility becomes relevant for safeguards when it handles fissile material or indulges in enrichment or reprocessing activities but none of these facilities was involved in any such activity. They have been declared civilian so that they can enter into international cooperation because all these facilities figured in "Entities List" of the US

Department of Commerce and were unable to collaborate with international partners.[78]

Lastly, the decision to shut down the CIRUS reactor and to move the fuel core of Apsara reactor out of the BARC and to place it under the safeguards was purely made to keep BARC off the sight of IAEA inspectors. Because both these reactors were located at BARC which is of great strategic importance. To quote Indian Prime Minister Manmohan Singh, "We have decided to take these steps rather than allow intrusive inspection in a nuclear facility of high security importance."[79]

HYDE ACT AND INDIAN CONCERNS

The "Henry J. Hyde United States-India Peaceful Atomic Energy Act of 2006" also called Hyde Act after the name of outgoing Chairman of the House International Relations Committee, Henry J. Hyde is a US domestic Law. The Act is a reconciled version of the House and Senate bills. The House version was passed on 26 July, 2006 by 356-68 votes and Senate version was passed on November 16, 2006 by 85-12 votes. Since these versions were not identical and as per US practice a bill cannot become law unless passed in identical form by US Congress. So, on December 9, 2006, the Joint Conference of two chambers of the US Congress passed a reconciled version of H. R 5682 and S. 3709 [80] which was signed into law by US President George W. Bush on December 18, 2006. The Act provides a legal basis for civil nuclear cooperation between India and US. It was necessary for Bush administration to enact a legislation waving the requirements of US Atomic Energy Act of 1954. As per the provisions of US Atomic Energy Act of 1954, a country (non- nuclear weapon state) becomes eligible for civil nuclear cooperation with US only when it fulfils certain requirements. These are:

- It should a Comprehensive Safeguards Agreement in place.
- It should not have exploded a nuclear device in past.
- It should not be involved in activities leading to the production of nuclear explosive devices.[81]

Since India does not fulfil any of these requirements so it was necessary for Bush administration to waive these requirements to engage India in civil nuclear cooperation and this was done by enacting Hyde Act. But the Act was drafted in such a way that it created serious concerns among Indian political and scientific establishments. The Act contains number of provisions which are not only contrary to July 18, 2005 understanding but also to the assurances given by Indian Prime Minister to Indian Parliament on number of occasions on civil nuclear cooperation with US. Following are some of the main objectionable provisions of the Act and the response given to them by Indian Prime Minister in Parliament.

1. Section 103 (a) (5) of the Act in the "Statement of Policy" lays down that:

"Given the special sensitivity of equipment and technologies related to the enrichment of uranium, the reprocessing of spent nuclear fuel and the production of heavy water, work with the members of the NSG, individually and collectively, to further restrict the transfers of such equipment and technologies, including to India."[82]

This section denies "full civil nuclear energy cooperation" to India earlier guaranteed under July 18, 2005 joint statement which describes India as "responsible state with advanced nuclear technology" and states that India should acquire same benefits and advantages as other such states. July 18, 2005 joint statement reiterates that US will work with her close friends and allies to ensure full civil nuclear cooperation with India as it realizes the goals of promoting nuclear power and achieving energy security.[83] Indian Prime Minister reacted to this cooperation denial effort in these words:

> "We seek the removal of restrictions on all aspects of cooperation and technology transfers pertaining to civil nuclear energy ranging from nuclear fuel, nuclear reactors to reprocessing spent fuel i.e. full aspects of complete nuclear fuel cycle... We will not agree to any dilution that would prevent us from securing the benefits of full civil nuclear cooperation..."[84]

2. Section 102 (13) along with section 103(a) (6) of the Act negates fuel supply assurances to India earlier guaranteed in March 2, 2006 "separation plan". Section 102(13) states that:

"The United States should not seek to facilitate or encourage the continuation of nuclear exports to India by any other party if such exports are terminated under United States law."[85]

Same policy has been maintained in section 103 (a) (6) which states that it shall be the policy of United States to:

> "Seek to prevent the transfer to a country of nuclear equipment, material or technology from other participating governments in the NSG or form any other source if nuclear transfers to that country are suspended or terminated pursuant to this title, the Atomic energy Act of 19549 42U.S.C.2011 et seq.), or any other United States law."[86]

This policy is contrary to the March 2, 2006 "separation plan" mutually agreed between two sides. The plan underlines multilayered fuel supply guarantees for civil nuclear reactors placed under safeguards. It states that US will help India to build up a "strategic reserve" of nuclear fuel to guard against disruption of fuel and despite this if fuel supplies are disrupted, US and India will convene a group of friendly countries to restore fuel supplies to India.[87] These assurances were repeated by Indian Prime Minster in the Parliament:

> *"Separation plan includes fuel supply assurances given by United States. Understandings in the separation plan also provide for contingency of fuel supplies to India... In the event of disruption of fuel supplies despite the assurances, India will have a right to take corrective measures to ensure the operation of its reactors."*[88]

3. Section 103 (b) of the Act stipulates that with respect to South Asia it shall be the policy of United States to:

"(1) Achieve at the earliest possible date, a moratorium on the production of fissile material for nuclear explosive purposes by India, Pakistan and the people's Republic of China.

(2) Achieve at the earliest possible date the conclusion and implementation of a treaty banning the production of fissile material for nuclear weapons to which both the United States and India become parties.

(3) Secure India's -

(A) full participation in the Proliferation Security Initiative;

(B) formal commitment to the statement of interdiction principles of such initiative;

(C) public announcement of its decision to conform its export control laws, regulations and policies with the Australia Group and with the guidelines, procedures, criteria and Control List of the Wassenaar Arrangement..."[89]

It has been India's long standing policy to seek universal nuclear disarmament along the lines of Rajiv Gandhi Action Plan and to promote the cause of non-discriminatory, general and complete disarmament but through the Act US wants to seek Indian support to achieve its regional non-proliferation goals and to bring India within the ambit of NPT notwithstanding the fact that India considers NPT regime as discriminatory and has always resisted the past proposals of regional disarmament. So the Act is seen as an effort on the part of US to derail Indian efforts of universal nuclear disarmament and to keep them confined to regional disarmament only.[90] This policy was touched up on Indian Prime Minster in these words:

> *"Our position on this matter is unambiguous. We are not willing to accept a moratorium on the production of fissile material. We are only committed to negotiate a Fissile Material Cut off Treaty in the Conference on Disarmament in Geneva, a commitment which was given by the previous government....Our commitment towards non-discriminatory global nuclear order remain unwavering in line with Rajiv Gandhi Action Plan. There is no dilution on this count. We do not accept proposals put forward from time to time for regional non-proliferation or regional disarmament."*[91]

On the Proliferation Security Initiative, PM said that it is an extraneous issue and is beyond the parameters of

July 18 understanding, therefore, India cannot accept it as a condition to implement the July 18, statement.

4. Section 103 (b) (4) of the Act requires US to:

"Secure India's full and active participation in the United States efforts to dissuade, isolate and if necessary, sanction and contain Iran for its efforts to acquire weapons of mass destruction including a nuclear weapons capability and the capability to enrich uranium or reprocess nuclear fuel and the means to deliver weapons of mass destruction."[92]

Reacting to this policy PM told Rajya Sabha that:

> "the proposed US legislation on nuclear cooperation will not be allowed to compromise India's sovereignty. Our foreign policy is determined solely by our national interests. No legislation enacted in a foreign country can take away from us that sovereign right... Our sole guiding principle in regard to our foreign policy whether it is Iran or any other country will be dictated entirely by our national interests."[93]

5. Section 104 (g) (2) of the Act calls upon US President to submit annually to appropriate Congressional Committees a detailed report on the measures undertaken by government of India to support US non-proliferation efforts particularly her efforts to contain Iran's nuclear programme and under section 104 (g) (2) (h), the President has to submit cyclically a report to different committees of US Congress on different aspects of India's nuclear programme including:

(i) The amount of uranium mined and milled during previous year.

(ii) The amount of uranium used or allocated for the production of nuclear weapons.

(iii) The rate of production of fissile material to be used for the production of nuclear weapons.

(iv) The rate of production of nuclear weapons in India.[94]

In response to these provisions PM stated that:

"we have made it clear to the United States our opposition to these provisions even if they are projected as nonbinding on India, as being contrary to the letter and spirit of the July Statement. We have told the US Administration that the effect of such certification will be to diminish a permanent waiver authority into annual one."[95]

He further added that:

"We have made it clear to the US that India's strategic programme is totally outside the purview of the July Statement and we oppose any legislative provisions that mandate scrutiny of either our nuclear programme or our unsafeguarded nuclear facilities."[96]

5. Section 104(5) subsection (iii) of the Act lays down that:

"In the event the IAEA is unable to implement safeguards as required by an agreement for cooperation arranged pursuant to section 123 of the Atomic Energy Act of 1054(42U.S.C.2153), appropriate assurance that arrangements will be put in place expeditiously that are consistent with the requirements of section 123 a(1) of such Act(42 U.S.C.2153(a)(1) regarding the maintenance of safeguards as set forth in the agreement regardless of whether the agreement is terminated or suspended for any reason."[97]

In the "separation plan" it was agreed that India will negotiate an India Specific Safeguards Agreement with IAEA and will place its civilian reactors under safeguards but will retain the right to take "corrective measures" (right to withdraw reactors from safeguards) if fuel supply is disrupted. But under Hyde Act US has to ensure that India's civil nuclear reactors remain under safeguards permanently notwithstanding the termination or suspension of the Agreement. Thus US inspectors will take over the responsibility of IAEA if for any reason it fails to keep watch over the Indian nuclear facilities. Reacting to this situation Indian Prime Minister stated that:

> "We will not accept any verification measures regarding our safeguarded nuclear facilities beyond those contained in an India specific Safeguards Agreement with IAEA. Therefore, there is no question of allowing American inspectors to roam around our nuclear facilities."[98]

6. Section 106 of the Act states that:

"A determination and any waiver under section 104 shall cease to be effective if the president determines that India has detonated a nuclear device after the enactment of this title."[99]

Indian Prime Minister responded in these words:

> "There is a provision in proposed US law that were India to detonate a nuclear explosive device, the US will have the right to cease further cooperation. Our position on this is unambiguous. The US has been intimated that the reference to nuclear detonation in the India-US bilateral Nuclear Cooperation Agreement as a condition for cooperation is not acceptable to us. We are not prepared to go beyond a unilateral voluntary moratorium on nuclear testing... India's possession and development of nuclear weapons is an integral

part of our national security. This will remain so."[100]

HYDE ACT VERSUS 123 AGREEMENT

The safe passage of the Hyde Act in the US Congress without accommodating Indian concerns generated a storm of controversy in India. It sent a clear message to the Indian political elite the Act has been passed with a definite purpose: to achieve broad strategic goals in the South Asian region. The passage of the Act not only jeopardised the fate of civil nuclear deal but also set in motion a hot debate involving political elite, academicians and strategic experts regarding the implications of Hyde Act for 123 Agreement. The UPA-I (United Progressive Alliance) government adopted the position that the Hyde Act is only a US domestic law that is binding on US only, India's rights and obligations regarding civil nuclear cooperation flow from 123 Agreement only[101] and 123 Agreement being an international agreement is independent of the Hyde Act because as per international legal standards an international agreement cannot be subjected to a national law.[102] It is true that Hyde Act is only an enabling legislation not binding on India but it binds the other party of the Agreement i.e. US which is clearly laid down in Article 2 of the 123 Agreement which states that "Each party shall implement the Agreement in accordance with its respective applicable treaties, national laws, regulations and license requirements concerning the use of nuclear energy for peaceful purposes."[103] So, US will implement the Agreement as per its national laws and applicable treaties¯NPT, Atomic Energy Act of 1954, Nuclear Non-Proliferation Act of 1978 and Hyde Act of 2006. This was also confirmed by US Undersecretary Nicholas Burns that the Agreement will be implemented in a manner consistent with Hyde Act. A group of Indian legal experts has even maintained that the Agreement is subject to internal laws of both the countries as well as to the amendments carried out in these laws and to any law that is enacted in future.[104] Since India does not have a comprehensive law governing civil nuclear cooperation with other countries, so the Agreement will be governed by US national laws. Past experience also reveals that US gives due regard to its national laws in the implementation of the 123 Agreements. In the case of 1963 Agreement signed with India, US invoked Nuclear Non-Proliferation Act of 1978 to deny fuel supply to India in the wake of her 1974 nuclear test. In contrast to this, Indian government claims that the 123 Agreement is an international treaty and it will be governed by the provisions of international law as the Agreement lays down in Article 16.4 that "the Agreement shall be implemented in good faith and in accordance with the principles of international law" which is a clear reference to the Vienna Convention on the Law of Treaties. Article 27 of the Vienna Convention states that "A party may not invoke the provisions of its internal law as a justification for its failure to perform a treaty."[105] Moreover, Article VI (2) of the US constitution states that the obligations of an international treaty override the provisions of a domestic law. This was also noted by then Indian Finance Minister P. Chidambaram in his speech in Lok Sabha:

> *"Under Article VI (2) of the US constitution all treaties made or which shall be made under the authority of United States shall be the supreme law of the land. In any view of the matter, the Hyde Act which is a domestic law, cannot bind India and cannot interfere with the implementation of the 123 agreement, which when ratified by the US Congress will be a bilateral treaty between two sovereign countries."*[106]

This view is also shared by the figures like Sean Murphy and Fredric L. Kirgis. Sean Murphy, Professor of Law at George Washington Law School holds that "An international agreement like 123 Agreement if approved by the Congress does take precedence over an earlier statue."[107] Similarly Fredric L. Kirgis, a constitutional diplomacy expert claims that an agreement reached between two sovereign countries supersedes the internal laws of either country.[108]

With reference to the above arguments following points are noteworthy:

- In the first place, Art. 27 of the Vienna Convention is not applicable to Indo-US 123 Agreement because Article 31 of the Convention states that the 123 Agreement shall be interpreted in the way it is written and Article 2.1 of the Indo-US Agreement explicitly states that it shall be implemented according to national laws and treaties,[109] so it will be implemented in that spirit.
- Secondly, US has signed the Vienna Convention on the Law of Treaties but has not ratified it, so it is not bound by it. More importantly US has a past record of not giving due regard to international treaties.
- Thirdly, US does not accord 123 Agreements the status of a treaty. It considers such Agreements only framework agreements which provide authorisation for cooperation in the field of civil nuclear energy and don't have a binding nature. This was confirmed by former US Ambassador to India David Mulford: "123 Agreement is a law but not everything in it is binding. This is the way global agreements are made."[110]

Hyde Act and 123 Agreement: Consistent or Inconsistent.

It is clear that 123 Agreement is not independent of the Hyde Act. The Act will definitely have a bearing on the implementation of the Agreement. Also, being a framework agreement, 123 Agreement cannot supersede the provisions of the Hyde Act. A careful analysis of the 123 Agreement and Hyde Act also drives home the point that there is no contradiction between the two and they are fully consistent with each other. Had there been any contradiction between the two, then 123 Agreement would not have sailed through US Congress. This was testified by Condoleezza Rice before the House Foreign Affairs Committee:

> *"We will support nothing with India that is in contradiction to the Hyde Act. It will have to be completely consistent with the obligations of the Hyde Act. We'll have to be consistent with the Hyde Act or I don't believe we can count on Congress to make the next step."*[111]

So, the Agreement was negotiated while keeping in mind the provisions of the Hyde Act. It is in full conformity with the Hyde Act and does not address any of the concerns raised by the Hyde Act.

In the first place, 123 Agreement does not guarantee full nuclear fuel cycle to India denied by the Hyde Act under section 103(a) (5). Article 5.2 of the Agreement states that heavy water production technology and sensitive nuclear facilities may be transferred to India pursuant to an amendment to the Agreement. Regarding the transfer of dual use items that could be used in enrichment, reprocessing or heavy water production facilities the Agreement states that such transfers will be governed by applicable laws, regulations and license policies of the both the parties.[112] So, the full civil nuclear cooperation with India is only possible when Agreement is amended, but the US State Department has made it clear that it has no intention to amend the Agreement,[113] because US does not consider such transfers necessary for the nations who seek to use nuclear energy for peaceful purposes. Such technologies are more used for weapons purposes and it is the policy of United States not to transfer such technology to any state including India. This was testified by Undersecretaries Nicholas Burns and Robert Joseph before the Senate Committee that "full civil nuclear cooperation [with India] will not include enrichment or reprocessing technology... Administration was not contemplating any transfers of those relating to production of heavy water either."[114] Agreement makes the transfer of dual use items subject to parties respective applicable laws and regulations, so nothing substantial in this matter has been achieved either, because it is the policy of United States not to transfer such items for the use in sensitive nuclear facilities.

Secondly, the Agreement makes "fuel supply assurances" conditional upon the approval of US Congress. In Article 5.6 (a) the Agreement notes that "the United States has conveyed its commitment to the reliable supply of fuel to India. Consistent with July 18, 2005, Joint Statement, the United States has also reaffirmed its assurance to create necessary conditions for India to have assured and full access to fuel for its reactors." But the same Article makes the fuel supply subject to amendment of US domestic laws as it states, "As part of its implementation of the July 18 , 2005 Joint Statement the United States is committed to seeking agreement from US Congress to amend its domestic laws and to work with friends and allies to adjust the practices of the Nuclear suppliers Group to create necessary conditions for India to obtain full access to international fuel market, including reliable, uninterrupted and continued access to fuel supplies from firms in several nations," [115] so the administration will have to go back to US Congress to amend its domestic laws to ensure uninterrupted fuel supplies to India which is a remote possibility.

Article 5.6 (b) of the Agreement states that the US will help India to build strategic reserve of fuel to guard against any disruption of fuel and despite this if any disruption in fuel supply occurs, the two countries will jointly convene a group of supplier countries to restore fuel supply to India. But US interpretation of this article is somewhat different from Indian one. US is committed to restore fuel supplies to India only if disruption occurs due to market failures and not due to India's own faults like detonation of nuclear device. This was confirmed by Robert Joseph and Nicholas Burns before the Senate Committee on Foreign Relations:

> *"Our negotiators were very clear that, while the US would be willing to provide reasonable fuel assurances designed to counter market imperfections, fuel assurances are not a 'condition' to any of India's commitments under the plan—including, in particular, safeguards in Perpetuity."*[116]

The Joint Conference Report of the two chambers of the US Congress on Hyde Act also explicitly makes it clear:

> *"The conferees understand and expect that such assurance[s]... will be concerned with disruption of fuel due to market failures or similar reason and not due to actions that are inconsistent with July 18, 2005 commitments, such as nuclear explosive test."*[117]

Besides this, the Agreement also does not make it obligatory upon US to help India to build up stockpile of nuclear fuel, but under the Agreement US is obliged to provide India only that amount of fuel sufficient for "efficient and continuous" operation of her safeguarded reactors (Article5.4).

Thirdly, the provision of "special verification visits" of Indian nuclear facilities to be conducted by US inspectors as mandated under section 104(d)(5)(B) (i) and (iii) of the Hyde Act has not been eliminated and is intact in 123 Agreement. Article 10, paragraph 4 of the Agreement states that if IAEA is not in a position to apply safeguards then parties will agree up on "appropriate verification measures" to ensure that safeguarded material is not diverted.[118] Same position is also maintained under Article 16.3 which states that safeguards have to be maintained notwithstanding the termination/ suspension of the Agreement. So when read with paragraph 4 of Article 10 it makes it clear that US material will always remain under safeguards whether performed by IAEA or USA. During a debate on the Hyde Act Senator Biden noted that:

> "Indian officials are reportedly upset that American personnel might need to visit India's nuclear sites. It should come as no surprise, however, that we need to ensure US nuclear material, equipment and technology are not diverted to military uses."[119]

Furthermore, the Agreement also provides for end-use monitoring under Article 12(3) and there by meets the requirements of section 104(d) (5) (B) (i) of the Hyde Act by stating that where cooperation "requires visits of experts, the parties shall facilitate entry of experts to their territory and their stay there in consistent with national laws, regulations and practices."[120]

Lastly, the Agreement does little to remove constraints on India's nuclear weapons programme particularly on her right to test a nuclear device. The Agreement makes no explicit reference to India's right to test a nuclear device. So the matters on which Agreement is silent US national laws will dictate terms. Section 123 of US Atomic Energy Act of 1954 states that if the partner country explodes a nuclear device for any reason, nuclear cooperation with that country will come to an end and that the partner country has to return all the nuclear material and equipments received from US.[38] So, if India conducts a nuclear test in future, US will terminate the Agreement and will cease its nuclear cooperation with India. The only relief that Agreement grants to India is that before terminating the Agreement, it calls upon the parties to hold consultations and take into account the relevant factors which prompted India to conduct a nuclear test (Article 14.2).

DEAL AND THE OPPOSITION

Indo-US Civil Nuclear Cooperation Agreement ran into trouble since the day it was announced as it faced a stiff opposition in both the countries. In spite of its strategic importance, the deal witnessed a severe opposition from nuclear non-proliferationists, strategic and political analysts, political and scientific communities and even from some sections of media.

OPPOSITION IN UNITED STATES

In US deal faced an up-front opposition from nuclear non-proliferation lobby which left no stone unturned to see the Agreement voted down by US Congress. In 2008, 150 non-proliferation specialists and non-governmental organisations addressed a letter to Foreign Ministers of NSG expressing serious concerns regarding the Agreement. The letter underlined that India has not signed NPT and has refused to halt its production of fissile material and nuclear arsenal but in spite of this has been allowed to receive the benefits of civil nuclear technology which is only a privilege of NPT signatory countries. This could create two groups of proliferates- "good" and "bad" which could prove dangerous for non-proliferation regime.[121] The charge levelled against the Agreement by non-proliferation priests was that it undermines Nuclear Non-proliferation Treaty which obligates every state party to the treaty not to assist any country either directly or indirectly in the acquisition or manufacture of nuclear weapons or nuclear explosive material. Since India has not signed NPT and has an active nuclear weapons programme but in spite of this has been blessed by US with a Civil Nuclear Cooperation Agreement which would help her to make quantative improvements in her nuclear weapons stock. This made Henry Sokolski to say that by signing nuclear deal with India US will join the ranks of North Korea and Iran as NPT violator. Strobe Talbott, former US Deputy Secretary of State termed the deal as the "breakdown" of NPT regime. He charged the Bush Administration for giving India more privileges than NPT states and said all this could have an adverse affect on the behaviour of nuclear have-nots and they will treat "NPT as anachronism, reconsider their self restraint and be tempted by the precedent that India has successfully established."[122]

In his Testimony before the Senate Committee on Foreign Relations, Robert Gallucci, Dean of the Edmund Walsh School of Foreign Service submitted that the deal would legitimize India's nuclear weapons programme and would grant India the status of a nuclear weapon state. He further said that by signing the Civil Nuclear Cooperation Agreement with India "we will put at risk a world of very few nuclear weapons states, and open the door to true proliferation of nuclear weapons in the years ahead."[123] Howard Berman, member of the House of Representatives said that the deal will make it difficult for US to stymie the spread of the weapons of mass destruction because "once you change the long-established non-proliferation rules for the benefit of one country...then it becomes much easier for other countries to justify carve-outs for their special friends."[124] In his opening statement on US–India Civil Nuclear Cooperation Agreement before the Senate Committee on Foreign Relations, Senator Richard G. Lugar of Indiana stated that in spite of India's commitment to bring more reactors under the safeguards, the Agreement would do little to prevent the further expansion of India's nuclear arsenal. He further said that "If Congress approves this agreement, we will be establishing a new course after decades of declining any cooperation with India's nuclear programme."[125]

Non-proliferationists also argued that the Agreement will weaken the case of US against rogue states like Iran and North Korea and would encourage other states to offer similar deals to their allies. For this reason, Senator Byron Dorgan called the deal a "grievous mistake that would reward rogue behaviour."[126] George Perkovich, a strategic analyst claimed that the Agreement "would set a precedent for China, a major nuclear exporter and an alleged source of Pakistan's nuclear technology to seek similar rule changes or exception."[127] Similar argument was put forth by Leonard S. Specter, Deputy Director of Centre for Non-Proliferation International Relations and Security Network, who said that:

> *"If you allow India to achieve same facilities as legitimate nuclear states signed the nuclear non-proliferation treaty (NPT), many other nations are likely to tread through it. China is before now believing of selling other reactors to Pakistan."*[128]

Senator Tom Harkin said that the Agreement would set an example of defying the most important non-proliferation instrument ˉ the Nuclear Non-Proliferation Treaty.[129] Aston B Carter held that the deal:

> *"would create a serious blow to non-proliferation regime and could set a dangerous precedent for rogue countries like Iran, North Korea and other aspiring nuclear powers... The deal has given a message that forgiveness comes to proliferators who wait long enough and that nuclear aspirants could bypass the NPT, if they waited long."*[130]

In an appearance before the Senate Committee on Foreign Relations, he further claimed that nuclear power can do little to meet India's future demand of energy. In future, India's energy scenario will be dominated by coal. So, burning this source "cheaply and cleanly" will do more to address India's energy and environmental concerns.[131]

The deal was also criticised on the grounds that neither it places any restriction on India's nuclear weapons programme nor obligates India to stop the production of fissile material has other nuclear powers have done. Contrary to this, deal gives India access to international fuel market which would free India's indigenous reserves of uranium which could be used for the production of nuclear weapons. Making a similar argument during a hearing on the "United States and India Nuclear Promotion Act of 2006" in the House of Representatives, Ed Markey, member of the House of Representatives from Massachusetts pleaded that the Agreement would free India's domestic reserves of uranium for the production of nuclear weapons and this could trigger a nuclear arms race in Southeast Asia which ultimately will spread across the world and urged the members of the House to vote down the proposal.[132] Critics also challenged the credibility of India "separation plan" on the non-proliferation grounds. Rejecting the non-proliferation gains of the "separation plan" Gary Milhollin, Director of the Wisconsin Project on Nuclear Arms Control said that:

> *"India's offer [separation plan] is like that of a counterfeiter with a 22 room house, who offers to let the police look into 14 rooms as long as they stay out of all others. Why would any policeman in his right mind accept such an offer... It would waste his time, just as it will waste the time of international inspectors to look at India's 14 declared reactors. Everyone knows that it will be eight undeclared ones that make the bombs."*[133]

OPPOSITION IN INDIA

In India deal also witnessed a widespread opposition from certain political and scientific communities. The opposition was so vibrant that government was forced to face a trust vote in Indian Parliament. The main political parties involved in the opposition were BJP and the left front. BJP, the main opposition party in the Parliament criticised the deal on the grounds that it undermines India's independent foreign policy and autonomy of her strategic nuclear programme. In a joint statement senior BJP leaders stressed that the deal is unacceptable to them as it compromises India's long term strategic programme. Yashwant Sinha and Arun Shourie, two senior leaders of the party termed the deal as an "assault on India's sovereignty and foreign policy" and demanded that the Agreement must be examined by Joint Parliamentary Committee and approved by the Parliament.[134] Speaking on the issue in Lok Sabha, former Indian Prime Minister Atal Bihari Vajpayee noted that by agreeing to bifurcate India's nuclear programme into military and civilian ones, the Agreement places an "indirect ban" on India's nuclear weapons capability. He further said that the Agreement does not recognise India as a Nuclear Weapon State (NWS), but only "A responsible state with advanced nuclear technology."[135] Criticising the government for its support to a West sponsored resolution at IAEA against Iran, Major General B. C. Khanduri observed that the vote was against national interests of the country and was exercised under US pressure. Recognising the importance of India-US strategic partnership, he noted that this relationship must be based on equal terms. India wants a long and enduring friendship with US but it must be subject to the limitations of our national interests, he further added.[136]

In his address to India conclave, leader of the opposition in Lok Sabha, L.K Advani stated that the nuclear deal was a "treaty between unequals" and it takes away India's sovereign right to test a nuclear device in future. He further said that:

> *"We are opposed to any agreement which says Indira Gandhi may have done Pokhran I, Atal Bihari Vajpayee may have done Pokhran II, but hereafter no government must do Pokhran III. If this is the price to be paid in the name of nuclear energy, our party is not in favour."*[137]

During a debate in Indian Parliament on the civil nuclear deal, Advani asserted that the deal jeopardises country's long term interests and BJP would renegotiate the deal if voted to power. The party was also upset with passage of the Hyde Act in US Congress. Attacking the provisions of the Act in Lok Sabha, Advani said that the Act seeks to "cap, then rollback and ultimately eliminate India's nuclear weapons capability" and if government moved ahead with this deal under the provisions of the Act, it will seriously compromise India's sovereignty and independence of foreign policy.[138]

Left parties opposed the Agreement on the grounds that it is a part of a wider strategic cooperation between India and US which could have serious implications for India's independent foreign policy and strategic autonomy. During a debate in Lok Sabha, Prabodh Panda of the Communist Party of India (CPI) observed that the Agreement has far reaching consequences for India's independent foreign policy and marks a deviation form India's policy of "Non-Alignment". With this Agreement, he said, India has become "something more than a major non-NATO ally of the US" and this is in sharp contradiction to India's Non-Alignment foreign policy, initiated by India's former Prime Minister Pandit Jawahar Lal Nehru. "The essence of Non-Alignment policy is anti-imperialism. But this agreement and the recognition of USA is that India is supposed to be a junior partner of USA" to achieve US strategic goals and this is against India's "nationally agreed and independent foreign policy." "Hence this joint statement is quite unfortunate and it is contrary to our national tradition."[139] Basudeb Acharya of the Communist Party of India (Marxist) (CPI (M)) challenged the rationale of the government to promote nuclear power as a solution to country's energy crises and said that nuclear power is "costliest among thermal and hydropower." Through the Agreement, US wants to revive its moribund nuclear industry by selling nuclear reactors and other equipments to India, he further said.[140]

Left parties also highly criticised India's vote against Iran at IAEA and argued that by voting with western powers at IAEA, India not only departed from the stand adopted by NAM counties at IAEA but also distracted from its long held position that Iran being an NPT signatory country has every right to peaceful uses of nuclear energy. C. K. Chandrappan of the CPI termed India's vote against Iran at IAEA a "betrayal" and argued that instead of voting with European powers, India should have abstained from voting to show the people of Iran that we are not "betraying the faith that they have bestowed in India all the time in the past."[141] Rupchand Pal of the CPI (M) claimed that by voting against Iran "India has committed a grave mistake. It is unacceptable to us. It is in violation of the commitment made by the Congress and the UPA [United Progressive Alliance] through the NCMP [National Common Minimum Programme] to the nation. This is the plan of America and we are just being a party to it and knowingly and unknowingly we are being put into a trap...Iran is pursuing its peaceful nuclear programme. America does not like it. America wants to re-dominate West Asia. They have a plan similar to the plan they had in Iraq and now, we, in India, are being a party to this game plan. It will be disastrous for a country which is known for its non-aligned role, for its role as independent and anti-imperialist throughout its history."[142]

CPI (M) challenged the claim of the government that nuclear deal will provide energy security to country. The party argued that nuclear energy is more expensive when compared with other sources of energy particularly with coal and hydro. It claimed that India's energy security lies in exploiting indigenous sources and ensuring future energy supplies from Iran and other countries in West and Central Asia.[143] General Secretary of the party, Prakash Karat argued that nuclear power will not make a substantial difference to country's power generation capacity and said that even if government succeeds to generate 20,000 MW of nuclear power by 2020 (target set by DAE) which is too ambitious and unrealistic, the share of nuclear power in country's power production would only rise to seven per cent from current three per cent which will be an insignificant contribution. Moreover the nuclear power is too much expensive. The cost of power generated from imported reactors will be twice that of coal fired power plants.[144]

Left parties also expressed deep concerns regarding various provisions of the Hyde Act. In a joint statement, the parties observed that the Act contains directions on India's foreign policy and strategic autonomy and requires India to have a foreign policy "congruent" to the foreign policy of US. Moreover, the Act takes precedence over the 123 Agreement and can be used to scrap to the 123 Agreement if India did not adhere to US foreign policy goals.[145] Providing a sharp reaction to the extraneous provisions of the Act on the floor of the Lok Sabha, Rupchand Pal noted that:

> *"The Hyde Act is unacceptable. It is violative of the assurances given by the hon. Prime Minister on the floor of the House. It is violative of the basic interests. It does not serve our long-term nuclear energy interest. It affects our sovereign, autonomous nuclear programme. It tries to bind us with its own foreign policy and philosophy at the cost of our independent foreign policy. It is unacceptable."* [146]

Not only the political establishments but India's nuclear scientific community was also alarmed by the passage of Hyde Act. In a note prepared by eminent nuclear scientists, it was underlined that:

- Act contains a number of provisions which are outside the scope of July 18, 2005 understanding.
- The Act requires India to work with US in achieving her regional non-proliferation goals and calls up on US president to determine annually that India is working with US in this regard.

- The Act contains provisions which could prove detrimental to India's strategic nuclear programme. It forbids India's future nuclear test and makes it clear that any such attempt would lead to termination of the Agreement and return of all nuclear material and equipment to US which would cause huge economic loss to the country.[147]

CONCLUSION

The Indo-US Civil Nuclear Cooperation Agreement which remained a hot topic for more than three years can be considered a great achievement for emerging strategic cooperation between two countries. With the initiative, two sides have buried their past animosities and have managed to remove a biggest irritant in their bilateral relationship. The Agreement which is being looked as a vital tool to forge a close strategic cooperation between the two countries favours India in many areas irrespective of the fact that India has not signed NPT. Inter alia, the Agreement has put an end to India's decades old nuclear isolation and gives India access to nuclear fuel and technology but falls short of Indian expectations. By barring the transfer of sensitive nuclear technology to India and linking the reliable fuel supplies to market failures US has changed the goal posts of full civil nuclear cooperation. More importantly by enacting Hyde Act which binds India in a number of crucial areas and links civil nuclear cooperation to India's good behaviour credentials, it seems that the civil nuclear cooperation is only cover under which US wants to achieve its broad strategic goals. Moreover, there is a difference in perception between two countries vis-à-vis Civil Nuclear Cooperation Agreement which can lead to a confrontation in future. Since the 123 Agreement does not have a provision to settle down the disputes arsing out of the interpretation or implementation of the Agreement by an arbitrational tribunal and only requires the parties to mutually settle down any such dispute, however, failure on the part of the parties to reach to an amicable solution can put India in a lose-lose situation.

REFERENCES

1 A G. Noorani, April 1981, "Indo-US. Nuclear Relations," *Asian Survey*, vol. 21, no.4, p. 401.

2 Paul F. Power, June 1979, "The Indo-American Nuclear Controversy," *Asian Survey*, vol. 19, no. 9, pp. 582, 596.

3 S. Paul Kapur and Summit Ganguly, 2007, "The Transformation of US-India Relations: An Explanation of Rapprochement and Prospects for Future," *Asian Survey*, vol. 47, no. 4, p. 647.

4 Sadia Tasleem, 2008, "Indo-U.S. Civil Nuclear Cooperation: Altering Strategic Positioning and Balance of Power in South Asia," *RSS Policy Studies* 44, pp. 22-23.

5 Quoted in Ajai K. Rai, 2009, *India's Nuclear Diplomacy after Pokhran II*, Delhi: Darling Kindersley, pp. 22-23

6 *Ibid.*, p. 22

7 *bid.* P. 21.

8 Cited in "India's Nuclear Debate: Indo-US Civil Nuclear Co-operation Agreement," 2008, *World Focus*, p. xi.

9 Quoted in Glenn Kessler, 2006, "India Nuclear Deal May face Hard Sell," *The Washington Post*, 4 April. Available at www.washingtonpost.com/wpdyun/content/article/2006/04/02/Ar2006040201315.html?referrer=email&referrer=email, (accessed on 22 April 2008).

10 Dhruva Jaishankar, 2009, "Chronicle of a Deal Foretold: Washington's Perspective on Negotiating the Indo-U.S. Nuclear Deal" in P. R. Chari (ed.), *Indo-U.S. civil Nuclear Deal: Seeking Synergy in Bilateralism*, New Delhi: Rutledge, p. 103

11 P. R. Lavoy, January/February, 2007, "Indian in 2006: A New Emphasis on Engagement," *Asian Survey*, vol. 47, no. 1, p.115.

12 Ajay K. Rai, *India's Nuclear Diplomacy after Pokhran II, op. cit.*, pp. 43-44.

13 Dinshaw Mistry, September/October 2006, "Diplomacy, Domestic Politics and U.S.-India Nuclear Agreement," *Asian survey*, vol. 46, no. 5, pp. 681-682.

14 Sadia Tasleem, "Indo-U.S. Civil Nuclear Cooperation: Altering Strategic Positioning and Balance of Power in South Asia," *op.cit.*, p. 27.

15 "Indo-U.S. Joint Statement of 18 July 2005." Available at http// www.meaindia.nic.in/mystart.php?id=53019889, (accessed on 5 June2012).

16 *Ibid.*

17 *Ibid.*

18 *Ibid.*

19 *Ibid.*

20 Quoted in John Cherian, March 10, 2006, "On a Mission to India," *Frontline*, vol. 23, issue 4.

21 M.R. Srinivasan, May 25, 2006, "Indo-US Civil Nuclear Agreement: Expectations and Concerns," *Current Science*, vol. 90, no. 10, p. 1318.

22 Quoted in Sadia Tasleem, "Indo-U.S. Civil Nuclear Cooperation: Altering Strategic Positioning and Balance of Power in South Asia," *op. cit.*, p. 35.

23 *Ibid.*, pp. 35-36.

24 Dhruva Jaishankar, "Chronicle of a Deal Foretold: Washington's Perspective on Negotiating the Indo-U.S. Nuclear Deal," *op. cit.*, pp. 116-117.

25 Quoted in K. Alan Kronstadt, 12 August, 2008, "India-US Relations," CRS Report for Congress, p. 3.

26 Siddharth Varadarajan, 2007, "Nuclear Deal Satisfactory: Kakodkar," *The Hindu*, 28 August.

27 *Ibid.*

28 Siddharth Varadarajan, 2007, "Kakodkar Supports Nuclear Deal," *The Hindu*, 28 July.

29 Quoted in "Bush Signs Nuclear Bill into Law," 2008, *The Hindu*, 10 October.

30 Siddharth Varadarajan, 2008, "Bush Signing Statement on 123 Leaves Flaws Intact," *The Hindu*, 10 October.

31 "Bush Signs Nuclear Bill into Law," *op. cit.*

32 Quoted in Siddharth Varadarajan, 2008, "India Reiterates 'legally Binding' Nature of 123 Pact's Provisions," *The Hindu*, 12 October.

33 Quoted in "India- United States Sign Historic Civil Nuclear Agreement," *India Review*, vol. 4, issue 11.

34 See the text of "Agreement for Cooperation Between the Government of India and the Government of United States of America Concerning Peaceful Uses of Nuclear Energy," p. 11

35 "This Agreement does not in any way Restrict our Strategic Autonomy," 2008, *The Hindu*, 14 September.

36 "Agreement for Cooperation Between the Government of India...," *op.cit.*, pp. 9-10.

37 Siddharth Varadarajan, 2007, "A Guide to the Provisions of 123 Agreement," *The Hindu*, 4 August.

38 Dr. Basudeo Prasad Yadav, October 2008, "Indo-US Civil Nuclear Cooperation Agreement," *Pratiyogta Darpan*, p. 641.

39 Quoted in Ajay K. Rai, *India's Nuclear Diplomacy after Pokhran II, op. cit.*, P .63.

40 John Cherian, February 25-March 10, 2006, "The China Factor," *Frontline* vol. 23, issue 4.

41 *Ibid.*

42 *Ibid.*

43 Quoted in Jabin T. Jacob, March 2006, "Indo-US Civil Nuclear Deal: The China Factor," *IPCS Special Report 14*, P.2.

44 Quoted in Jayshree Bajoria and Esther Pan, November 5, 2012, "The US-India Nuclear Deal." Available at http//www.crf.org/india/us-india-nucleardeal/19663, (accessed on 24 February 2012).

45 Quoted in Dr. Tom Sauer, May 2010, "Drivers behind the Nuclear Deal with India: US Domestic Politics or Geostrategic Concerns," p.11. Available at www.suc.kiley-ven-be/web/files/11/72wi6-115.pdf, (accessed on 25 May 2012.

46 See the text of "Henry J. Hyde United States-India Peaceful Atomic Energy Act of 2006."

47 K. Alan Kronstadt, "US-India Bilateral Agreements in 2005," *CRS Report for Congress.*

48 "Indo-US Deal Getting Bipartisan Support," 2006, *Dawn*, 7 April.

49 Quoted in Sadia Tasleem, "Indo-U.S. Civil Nuclear Cooperation: Altering Strategic Positioning and Balance of Power in South Asia," *op. cit.*, p. 79.

50 Quoted in Sudesh Rani, 2010, "Indo-US Nuclear Agreement: A Critical Study," *Punjab Journal of Politics*, vol. XXXIV, nos. 1-2, p. 78.

51 Quoted in Annapurna Nautiyal, September 2008, "Indo-US Civil

Nuclear Deal: What is There," *Strategic Insights*, vol. VII, issue 4.

52 Quoted in Prashant Hosur, Spring 2010, "The Indo-US Civilian Nuclear Agreement: What's the Big Deal," *International Journal*, p. 439.

53 Quoted in Dr. Tom Sauer, "Drivers behind the Nuclear Deal with India: US Domestic Politics or Geostrategic Concerns," *op. cit.*, p. 15.

54 *Ibid.*, p. 16.

55 "Remarks of Secretary of State Condoleezza Rice at the Senate Foreign Relations Committee on the US-India Civil Nuclear Cooperation Initiative," April 5, 2005, p. 7. Available at http// www.carnegieendowment.org/static/npp/reports/rice_4-5-06.pdf, (accessed on 22 July 2012).

56 A.P.J. Abdul Kalam and Srijan Pal Singh, 2011, "Nuclear Power is Our Gateway to a Prosperous Future," *The Hindu*, 6 November.

57 "Remarks of Secretary of State Condoleezza Rice at the Senate Foreign Relations Committee on the US-India Civil Nuclear Cooperation Initiative," *op. cit.*, p. 8.

58 K. R. Gupta, 2007, "Indo-US Nuclear Cooperation" in Rahul Bhonsle et al (eds.), *Indo-US Nuclear Deal* ,vol. II, New Delhi: Atlantic Publishers, p. 165.

59 Sudesh Rani, "Indo-US Nuclear Agreement: A Critical Study," *op. cit.*, p. 77.

60 "Remarks of Secretary of State Condoleezza Rice at the Senate Foreign Relations Committee on the US-India Civil Nuclear Cooperation Initiative," *op. cit.*, p. 10.

61 John Stephenson and Peter Tyman, March 2007, "Will the US-India Civil Nuclear Cooperation Initiative Light India" in Henry Sokoloski (ed.), *Gauging US-India Strategic Partnership*, P. 31. Report published by Strategic Studies Institute. Available at http/ /www.strategicstudiesinstitute.army.mil, (accessed on 5 April 2012).

62 "Uranium Crises", December 21 - January 13, *Frontline* vol. 22, issue 27.

63 Quoted in Zia Mian et al, "Plutonium Production in India and US-India Nuclear Deal," in Henry Sokoloski, *Gauging US-India Strategic Partnership, op.cit.*, P. 107.

64 Sharon Squassoni, December 22, 2006, "India's Separation Plan: Issues and Views," *CRS Report for Congress*. p. 2.

65 *Ibid.*, p.13.

66 Ashwin Kumar, December 31 – January 6, 2006, "Indo-US Nuclear Deal: Separation of Civilian and Military Facilities," *Economic and Political weekly*, vol. 40, no. 53, p.5530.

67 "PM'S Statement In Parliament on Civil Nuclear Energy Cooperation With United States," February 27, 2006. Available at: http//pmindia.nic.in/pmsinparliament.php?nodied=28, (accessed on 25 February2012).

68 Dinshaw Mistry, "Diplomacy, Domestic politics and US-India Nuclear Agreement," *op.cit.*, pp. 690-91.

69 *Ibid.*, pp. 686-687.

70 Sharon Squassoni, March 3, 2006, "India's Nuclear Separation Plan: Issues and Views," *CRS Report for Congress.*

71 "PM's Suo-Motu Statement on Discussions on Civil Nuclear Energy Cooperation with US: Implementation of India's Separation Plan," March 7, 2006. Available at: http//pmindia.nic.in/pmsinparliament.php?nodied=24, (accesseu on 25 February 2012).

72 "Implementation of The India-United States Joint Statement of July 18, 2005: India's Separation Plan."Available at http//www.dae.gov.In/pres/sepplan.Pdf, (accessed on 19 March 2012).

73 *Ibid.*

74 Quoted in Sharon Squassoni, "India's Nuclear separation plan: Issues and Views," *op. cit.*, p. 20.

75 Praful Bidwai, March 11 -24, 2006, "Nuclear Deal at What Price?," *Frontline*, vol. 23, issue 5.

76 PR Chari, March 2006, "Parsing the Separation Plan: The Indo-US Subsidiary Deal," *IPCS Issue Brief*, no. 36, pp. 2-3.

77 Siddharth Varadarajan, 2006, "Was Bush Speech a Warning on Separation," *The Hindu*,1 March.

78 M.R Srinivasan, May 25, 2006, "Indo-US Nuclear Agreement: Expectations and Concerns," *Current Science*, vol. 90, no. 10, pp. 1318-1319.

79 Quoted in "No Curb on Nuclear Facilities: Manmohan," 2006, *The Hindu*, 8 March.

80 R. Ramachandran, December 16-29, 2006, "Shifting of Goalposts," *Frontline*, vol. 23, issue 25.

81 "123 Agreement, Hyde Act and Indian Predicament," 2008, *Current Affairs*, 5 March.

82 See the text of the "Henry J. Hyde United States-India Peaceful Atomic Energy Act of 2006," *op. cit.*

83 See "Indo-US joint Statement of 18 July, 2005," *op. cit.*

84 "Statement of PM in Rajya Sabha on India-US Nuclear Agreement," August 17, 2006. Available at http// pmindia.nic.in/ pmsinparliament.php?nodied=27, (accessed on 12 June 2012).

85 See "Henry J. Hyde United States-India Peaceful Atomic Energy Act Of 2006", *op.cit.*

86 *Ibid.*

87 "Implementation of the India-United States Joint Statement of July 18, 2005: India's Separation Plan,"*op. cit.*

88 Quoted in B. B. Singh, 2007, "The Hyde Act 2006: India's Nuclear Dilemma", *Atoms for Peace: An International Journal,* vol. 1, no. 4, p. 310.

89 See "Henry J. Hyde United States-India Peaceful...," *op.cit.*

90 M.R. Srinivasan, 2006, "India May Lose Control of Its Nuclear Future," *The Hindu*, 14 December.

91 "Statement of PM in Rajya Sabha on India-US Nuclear Agreement," *op. cit.*

92 See "Henry J. Hyde United States-India Peaceful... ," *op.cit.*

93 "Statement of PM in Rajya Sabha on India-US Nuclear Agreement," *op. cit.*

94 See "Henry J. Hyde United States-India Peaceful...," *op. cit.*

95 "Statement of PM in Rajya Sabha on India-US Nuclear Agreement," *op. cit.*

96 *Ibid.*

97 See "Henry J. Hyde United States-India Peaceful...," *op. cit.*

98 "Statement of PM in Rajya Sabha on India-US Nuclear Agreement," *op.cit.*

99 See "Hennery J. Hyde United States-India Peaceful...," *op.cit.*

100 "Statement of PM in Rajya Sabha on India-US Nuclear Agreement," *op.cit.*

101 Dr. Chintamini Mahapatra, March 10, 2008, "Hyde Act Controversy: Is it really Pertinent." Available at http// www.sarkaritel.com/news and features/infa/march2008/ 10hyde_contronersy.htm, (accessed on 20 March 2012).

102 "123 Agreement is Independent of Hyde Act: Congress," August 20, 2007. Available at http//news.oneidea.in/2007/08/20/123-

agreement-is-independent-of-hyde-act-congress-1872620790.html, accessed on 20 March 2012.

103 See the text of "Agreement for Cooperation Between the Government of India And ...", *op. cit.*, p. 5.

104 "Executive Cannot Act Independently on Parliament on Nuclear Deal," 2007, A statement issued by V. R. Krishna Iyengar, justice P. B. Sawant and Justice H. Suresh, *The Hindu*, 13 September.

105 Siddharth Varadarajan, 2007, "Deal Breather not Deal Breaker," *The Hindu*, 20 August.

106 "Hyde Act, 123 Agreement and India," Excerpts of P. Chidambaram's speech on the Motion of Confidence in the Lok Sabha. Available at: http//www.sarrijan.com/discuss/34-options/80-hyde-act-123-agreement-and-india, (accessed on 25 March 2012).

107 "Tactical Hold', 2007, *India Today*, 3 September.

108 *Ibid.*

109 "Left Stand on the Nuclear Deal: Notes Exchanged in the UPA-Left Committee on India-US Civil Nuclear Cooperation", July 2008, p. 125.

110 Quoted in Ajai k. Rai, *India's Nuclear Diplomacy after Pokhran II*, *op. cit.*, P. 65.

111 Quoted in Sandeep Dikshit, 2008, "Hyde Act is only a Domestic Legislation Says Boucher," *The Hindu*, 5 March.

112 See the text of "Agreement for Cooperation between The Government Of India...", *op. cit.*, p. 9.

113 "Questions for the Record submitted to Assistant Secretary Bergner by Chairman Tom Lantos House Committee on Foreign Affairs," October 5, 2007, Question No.6.

114 Arun Shourie, 2006, "Now Let the PM Square this Circle," *The Indian Express*, 28 November.

115 See the text of "Agreement for Cooperation between The Government of India...," *op. cit.*, p. 10.

116 Quoted in Arun Shourie, 2006, "Facts Versus Government's Fiction, "*The Indian Express*, 22 December.

117 Quoted in *ibid.*

118 Brahma Chellaney, 2007, "123: Text and context, Pact Unequal, US can Scrap it at Will," *Deccan Chronicle*, 4 August.

119 Quoted in Arun Shourie, 2006, "Time to Deal with Aftermath," *The Indian Express*, 30 November.

120 Cited in Brahma Chellaney, 2007, "Questions Raised by the 123 Agreement," *The Hindu*, 14 September.

121 Allison Marz Freedman, "USINAPC and U.S-India Nuclear Deal: Lasting Influence or One Short victory?," *CUREJ College Undergraduate Research Electronic Journal*, pp. 28-29.

122 "Good Day for India, Bad Day for Non-proliferation: Strobe Talbot," 2005, *Daily Times*, 25 July.

123 "U.S. Senator Richard G. Lugar Holds A Hearing on U.S.-India Atomic Cooperation," April 28, 2006 Political/Congress Transcript Wire.

124 Hearing on the "United States and India Nuclear Cooperation Promotion Act of 2006 (House of Representatives)," July 26, 2006.

125 Statement of Richard G. Lugar before the Senate Foreign Relations Committee during a Hearing on US-India Civil Nuclear Agreement, April 5, 2006.

126 Quoted in Sahib Gulati, "The US Domestic Debate on Indo-US Nuclear Deal Under Obama Administration: Implications for India." Available at http//www.observerindia.com/ccms/expert/orfonline/modules/analyss/attachments/implications-for-india-12393120213.pdf, (accessed on 3 April 2012).

127 Ajai k. Rai, *India's Nuclear Diplomacy after Pokhran II, op. cit.*, p. 95.

128 Quoted in Muhammad Ishaque Fani, "The Indo-US Strategic Partnership In Post 9/11: Implications For Pakistan," *Pakistan Vision*, vol.10, no. 2, p.143.

129 Sahib Gulati, "The US Domestic Debate on Indo-US Nuclear Deal under Obama Administration: Implications for India," *op. cit.*

130 Annapurna Nautiyal, "Indo-US Civil Nuclear Deal: What is There," *op. cit.*

131 Ashton B. Carter, November 2, 2005, "The Indian Deal: Looking at the Big Picture,"Testimony of the Ashton B. Carter before the Senate Committee on Foreign Relations.

132 Hearing on the United States and India Nuclear Cooperation Promotion Act of 2006 (House of Representatives), *op.cit.*

133 Testimony of Garry Milhollin, Director, Wisconsin Project on Nuclear Arms Control before the Senate Committee on Foreign Relations, April 26, 2006.

134 Neena Vyas, 2007, "123 Agreement an Assault on Nuclear Sovereignty: BJP," *The Hindu*, 5 August.

135 Lok Sabha Debates, August 3, 2005, vol. XI, no. 8, pp. 549-550.

136 Lok Sabha Debates, February 27, 2006, vol. XVII, no. 8, pp. 347-351.

137 Quoted in "Left Told Deal will not be Pursued: Advani," 2008, *The Hindu*, 5 March.

138 Lok Sabha Debates, December 18, 2006, vol. XXIV, no. 19, pp. 534-536.

139 Lok Sabha Debates, August 3, 2005, *op. cit.*, pp. 552-556.

140 Lok Sabha Debates, August 23, 2006, vol. XXII, no. 20, pp. 661-662.

141 Lok Sabha Debates, February 27, 2006, *op. cit.*, pp. 342-343.

142 *Ibid.*, p. 364.

143 Sandeep Dikshit, 2008, "Nuclear Energy is Cover, Intent Is Strategic US Ties: CPI (M)," *The Hindu*, 14 May.

144 Prakash Karat, July - September 2007, "The Indo-US nuclear Deal: Struggle to Defend National Security," *The Marxist*, vol. XXIII, no. 3.

145 See the "Statement of left Parties on Indo-Us Bilateral Nuclear Cooperation Agreement," August 7, 2007.

146 Lok Sabha Debates, December 18, 2006, *op. cit.*, p. 552.

147 "Hyde Act and Nuclear Scientists Note," 2006, *The Hindu*, 6 December.

3

The Deal at IAEA and NSG

INTRODUCTION

The July 18, 2005 Joint Statement underlines a commitment made by Indian Prime Minister Dr. Manmohan Singh that in order to realise the potential of full civil nuclear cooperation, India will negotiate a Safeguards Agreement with IAEA and will place its civilian reactors under the surveillance of IAEA. In compliance to this commitment, India started negotiations with IAEA and after several rounds of tough and hectic negotiations, two sides negotiated an India Specific Safeguards Agreement (ISSA) which was approved by IAEA Board of Governors on 1 August 2008. Though the Agreement was approved with consensus but its passage was not altogether a cakewalk. During the deliberations, number of countries expressed concerns regarding the Agreement and issued critical statements. The Agreement also became a subject of hot debate in both the countries due to some controversial issues linked to the Agreement.

After the approval of the Agreement by IAEA Board, US approached NSG with a proposal seeking adjustment of cartel's guidelines to exempt India from the requirement of Comprehensive Safeguards Agreement—a necessary condition for non-nuclear weapon states to engage in civil nuclear trade with rest of the world. Initially, US proposal faced a stiff resistance but ultimately US diplomatic might prevailed and India was granted a "waiver" by NSG but only after imposing some conditions on it.

This chapter tracks the process of negotiations at IAEA and NSG with a special focus on India-IAEA Safeguards Agreement and NSG "waiver" and analyses their implications for India.

INTERNATIONAL ATOMIC ENERGY AGENCY: A BRIEF INTRODUCTION

International Atomic Energy Agency is an international organisation established in 1957 to promote peaceful uses of atomic energy and to prevent its use for military purposes. IAEA was created in response to fears and expectations arising out of the discovery of nuclear energy. The genesis of the organisation can be traced from Acheson-Lilienthal Report of 1946 which proposed the establishment of an international body to control all nuclear activities which are dangerous for international community. It was also proposed that such an international body would "own, operate and develop the nuclear industry on the behalf of all the nations," would operate nuclear fuel fabrication plants and nuclear reactors, would carry out research in the field of nuclear energy including nuclear explosives and would be made the proprietor of nuclear ores and fuels.[1]

The Acheson-Lilienthal Report was presented by US delegate Bernard Baruch at the inaugural meeting of the

United Nations Atomic Energy Commission in June 1946 which came to be known as Baruch Plan. But the proposal was turned down by USSR on the grounds of national sovereignty. The proposal remained in cold waters till 1953 when it was revived by US President Dwight D. Eisenhower. In 1953, Eisenhower in his famous "Atoms for Peace" speech delivered at United Nations General Assembly said that in order to prevent the military use of atomic energy and to advance its use for peaceful purposes, the countries involved in the exploitation of atomic energy "should begin now and continue to make joint contributions from their stockpiles of normal uranium and fissionable materials to an international atomic energy agency." He further said:

> *"We would expect that such an agency would be set up under the aegis of United Nations... The atomic energy agency could be made responsible for the impounding storage and protection of the contributed fissionable and other materials... The more important responsibility of the atomic energy agency would be to devise methods whereby this fissionable material would be allocated to serve the peaceful pursuits of mankind."*[2]

However, the proposal of the Eisenhower could see the light of the day in 1956 when statute of International Atomic Energy Agency was approved at an international conference held at UN headquarters. The objectives of the Agency as laid down in the statue are to accelerate and enlarge the contribution of atomic energy to peace, health and prosperity throughout the world and to ensure that the assistance provided by the Agency or at its request or supervision is not diverted for military purposes.[3] Some of the important functions of the Agency are:

- To assist countries to carry out research on the application of atomic energy for peaceful purposes.
- To promote the exchange of scientific and technical information on the peaceful uses of atomic energy.
- To encourage the exchange of training of scientists and experts in the field of peaceful uses of atomic energy.
- To establish safeguards to ensure that the assistance provided by the Agency in the field of nuclear energy for peaceful purposes is not diverted towards military purposes and to apply safeguards when a request for that purpose is made by a state or states.[4]

IAEA HAS THREE MAIN ORGANS

1. Board of Governors

The Board of Governors (BOG) is one of the two policy making bodies of the Agency entrusted with the responsibility of formulating most of the policies of the Agency. The Board consists of thirty five members. Among these thirteen are designated by the outgoing Board for one year term, ten from the countries which are most advanced in the field of nuclear energy technology and three from the areas: North America, Latin America, Western Europe, Eastern Europe, Africa, Middle East and South Asia, South East Asia, the Pacific and the Far East. Remaining twenty-two members are elected by General Conference for a term of two years. Half of these members are elected each year. The Board meets five times in a year and performs some of the important functions like appointment of Director General and publishing of IAEA standards.[5]

2. General Conference

The General Conference is the highest policy making

body of IAEA. It is composed of all the members of the Agency. Currently, Agency has 151 members. General Conference serves a forum for debate on current issues and policies. It meets once in a year. However, Director General can convene a special session at the request of majority of the members or BOG. At every session, the members of General Conference elect a president to facilitate the smooth working of the General Conference. The president holds office for one session only. The General Conference approves the nominee for the post of Director General.[6]

3. Secretariat

The Secretariat is composed of professional, technical and other personnel staff appointed by the Director General who serves as the head of the Secretariat. The Director General is appointed by BOG with the approval of General Conference for a term of four years. Director General is Chief Administrative Officer of the Agency and is responsible for the enforcement of actions approved by the BOG and General Conference. Director General works under the authority of BOG and performs his duties in accordance with the rules and regulations adopted by the BOG. Present Director General is Yukiya Amano of Japan. He took office in December 2009.

IAEA SAFEGUARDS

Safeguards are a set of activities carried out by IAEA to verify that a state is living up to its commitments not to use its civilian nuclear programme for military purposes. The Agency derives its authority to implement and administer safeguards form Article III (A) (5) of its statute which confers on the Agency the authority to apply safeguards under two circumstances. First, when the Agency becomes the source of nuclear material, equipment, facilities, services or information and second, when a request to that purpose is made by a state unilaterally or by states parties to any bilateral or multilateral agreement.[7]

The objective behind the entire safeguards system is to ensure that nuclear material and activities placed under safeguards are not diverted for weapon purposes. Initially, the demand for safeguards was very less and principal suppliers used to perform this function themselves by maintaining supervision on the nuclear material and technology supplied to other countries. But as the trade grew and technology became more advanced, the suppliers found it difficult to maintain safeguards,[8] so the responsibility was shifted over to IAEA. It was in 1959 when Agency was first time called in to apply safeguards on natural uranium supplied by Canada to Japan. These safeguards were ad hoc in nature as the Agency had not developed a definite set of rules and regulations governing the application of safeguards. However, in 1961, a safeguards document INFCIRC/26 was published by the Agency which laid down the rules and regulations for the application of safeguards on small research and experimental reactors up to 100 MW. But due to advancement in nuclear trade and technology this system was revised and a new document INFCICR/66 was adopted in 1965 which extended safeguards to all sizes of reactors. This system was provisionally extended in 1966 (INFIRC/66 Rev.1) and in 1968 (INCIRC/66Rev.2) to cover reprocessing and fuel fabrication plants.[9] Under this system, states were at discretion to offer the nuclear material and facilities for safeguards and usually only imported material and facilities were offered for safeguards.

In 1967, Treaty for the Prohibition of Nuclear Weapons in Latin America and Caribbean (Tlatelolco Treaty) was successfully concluded which called upon the states parties to the treaty to conclude a Comprehensive Safeguards Agreement (CSA) with the Agency. Moreover, in 1968 Nuclear Non-Proliferation Treaty (NPT) was successfully negotiated which entered into force in 1970. The treaty made it obligatory upon all Non-Nuclear Weapon States (NNWS) parties to the treaty to place all their fissionable nuclear material and facilities under their jurisdiction under the safeguards of IAEA.[10]

In order to fulfil the requirement of NPT, a "safeguards committee" was set up by the IAEA Board of Governors to devise a framework for new safeguards applicable to all Non-Nuclear Weapon States, covering all their nuclear material and activities. The committee produced a document entitled "The Structure and Content of Agreements between the Agency and States Required in Connection with the Treaty on the Non-Proliferation of Nuclear Weapons".[11] It was approved by the Board in 1972 and was published as INFIRC/153 and became a model for the conclusion of Comprehensive Safeguards Agreements (CSAs) between the Agency and the NNWSs. This type of agreement makes it obligatory on a state to place all its nuclear material and facilities in peaceful nuclear activities under IAEA safeguards. Under this type of agreement IAEA has the authority to verify the "correctness" and "completeness" of a state's declarations. To put it simply, under the CSA, IAEA has the authority to verify that the information provided by a state regarding its declared nuclear activities is not only correct i.e. there is no diversion of nuclear material but also complete in the sense that there is no undeclared nuclear material or activities in the state. But the Agency was not well equipped to detect undeclared nuclear material and activities in a state and this loophole in the system came to light in 1991 when it came to the notice of the Agency that Iraq in spite of having a CSA in place was running a clandestine nuclear weapons programme. One year latter it was also revealed that North Korea which had acceded to the NPT and had signed a CSA with Agency was operating a secret reprocessing facility. These incidents triggered the need to strengthen the existing safeguards system and in this direction certain positive steps were taken. For instance, in 1992, Board affirmed that under CSA, Agency has the right to carry out special inspections not confined to nuclear facilities only but to any location within the territory of a state if Agency feels that more information is needed to draw "credible conclusions" regarding state's declarations or the concerned state has not provided all the relevant information to the Agency. It was also declared that the scope of Comprehensive Safeguards Agreements is not only confined to declared nuclear material but to all nuclear material that a state should have declared and in case of noncompliance, Agency can take recourse to United Nation Security Council to enforce the safeguards.[12] However, a major step in the direction of strengthening the safeguards system was taken in 1993 when "Programme 93+2" [13] was launched. Under this programme two sets of measures were identified to strengthen the system. The first set included measures which Board of Governors believed that could be implemented under the existing system. But the second set included measures which could not be implemented under the existing system but could be implemented only by conferring additional legal authority to the Agency. This responsibility was given to "93+2 Safeguards Committee" which prepared the text of Model Additional Protocol (MAP). MAP provided necessary legal authority to the Agency to enforce second set of measures.[14] The MAP was approved by IAEA Board on 15 May, 1997 and was published as IAEA document INFIRC/540. It arms the Agency with broader access rights and obligates a state to provide all the necessary information to the Agency regarding her nuclear fuel cycle activities. Inter alia, the MAP provides for:

- *"Information about and access to all aspects of a state's nuclear fuel cycle from uranium mines to nuclear waste and any other locations where nuclear material intended for non-nuclear uses is present;*
- *Short-notice inspector access to all buildings in a nuclear site;*
- *Information on the manufacture and export of sensitive nuclear-related technologies and inspection mechanism for manufacturing and import locations;*
- *Access to other nuclear related locations; and*
- *Collection of environmental samples beyond declared locations when declared necessary by the IAEA."*[15]

The purpose of all these measures is not only to empower the Agency to verify declared nuclear material and activates in a state but also to detect undeclared nuclear material, and activities.

TYPES OF SAFEGUARDS

IAEA, which is entrusted with the responsibility to prevent the diversion of nuclear material and activities has devised a well established system of safeguards over the years to prevent the use of nuclear material and technology for military purposes. Based on this system, safeguards agreements are concluded between the states and the Agency which serve as important instruments to prevent any sort of diversion. Currently, multiple types of safeguards agreements are in place concluded between the Agency and different states.

1. Comprehensive Safeguards Agreement (CSA)

This type of agreement is concluded between the Agency and a Non-Nuclear Weapon State party to Nuclear Non-Proliferation Treaty or any other regional non-proliferation arrangement like Tlatelolco Treaty, Rarotonga Treaty, Bangkok Treaty or any other bilateral treaty. Comprehensive Safeguards Agreements follow the structure and content of INFCIRC/153(corrected). Such agreements cover all "source or special fissionable nuclear material in all peaceful nuclear activities within the territory of the state, under its jurisdiction or carried out under its control anywhere."[16] Under CSA, the Agency has the authority to verify both declared and undeclared nuclear material and activities in a state but declared nuclear material, and activities remain focus of inspections. During inspections, inspectors mainly focus on "strategic points." These are those points which are very important for the implementation of safeguards.[17] Under such agreements, different types of inspections are carried out. These fall under three categories.

(i) Ad hoc inspections : These inspections are carried out to verify a "state's initial report on its nuclear material or reports on changes thereto and to verify nuclear material involved in international transfers."[18]

(ii) Routine inspections : These inspections are carried out frequently. These are made according to a definite schedule. Such inspections can also be carried out by serving a short notice to state or can be sometimes unannounced.

(iii) Special inspections : Such inspections are conducted if Agency believes that state is not providing adequate information to the Agency or if the Agency believes that more information is needed to draw credible conclusions regarding state's declarations.

Besides this, safeguards visits are also performed for verifying the design information of nuclear facilities provided by the state. Such visits are carried out throughout the lifetime of a facility i.e. form construction phase to decommissioning phase.[19]

2. Item Specific Safeguards Agreement

Under this type of agreement, specific number of facilities and material mentioned in the safeguards agreement are offered for safeguards. These agreements follow the structure and content of INFIRC/66/Rev.2. Currently this type of agreement is in force in three states which are not parties to the NPT¯ India, Pakistan and Israel.

3. Voluntary Offer Agreements (VOAs)

Under NPT, Nuclear Weapon States parties to the NPT are not obliged to enter into a safeguards agreement with IAEA, however, these states have voluntarily offered a limited number of nuclear facilities and material for safeguards. Such agreements are called Voluntary Offer Safeguards Agreements. These agreements follow the format of INFCIRC/153(corrected) i.e. which applies to CSAs. But unlike CSAs, under VOA a NWS places only such facilities and material under safeguards which are not important for strategic purposes and also enjoys the liberty to withdraw such material and facilities from safeguards when deemed necessary.

4. Additional Protocol

As mentioned earlier that after the revelation of Iraq's clandestine nuclear weapons programme, the Board of Governors approved "Model Protocol" additional to the agreements between States and the International Atomic Energy Agency for the application of safeguards in 1997 called Model Additional Protocol (MAP) as a measure to strengthen the safeguards system. So the states with a safeguards agreement in force are also encouraged to sign a protocol additional to their safeguards agreements. Additional Protocols follow the format of INFCIRC/540 (corrected). Model Additional Protocol enables the Agency not only to verify declared nuclear material and activities in a state but also to verify the absence of undeclared nuclear material and activities. However, it is not mandatory for a state to sign an Additional Protocol with the Agency along the lines of MAP. Under Model Additional Protocol, a state is obliged to provide much broader and wider information to the Agency covering all aspects of its nuclear fuel cycle activities, research and development, uranium mining, manufacture and export of items used in sensitive nuclear technologies, decommissioned facilities etc.[20] MAP arms the Agency with "complementary access" rights. "Complementary access" is a verification tool used by the Agency to fulfil three purposes:

- To ensure the absence of undeclared nuclear material and activities in a state.
- To confirm decommissioned status of nuclear facilities and other locations where nuclear material was used formerly.
- To resolve the questions and inconsistencies regarding the "correctness" and "completeness" of states declarations.

However, it must be noted that "complementary access" is not "access anywhere at any time." This tool is used by Agency selectively not systematically.[21]

INDIA–IAEA SAFEGUARDS AGREEMENT

India's need to conclude a Safeguards Agreement with IAEA stems from the commitment made by the Indian Prime Minister Dr. Manmohan Sing in July 2005 joint statement that India will conclude a Safeguards Agreement with the IAEA and will place her civilian nuclear facilities under IAEA safeguards. However, because of serious objections levelled against the Civil Nuclear Cooperation Agreement, government faced a stiff resistance from left parties to start negotiations with IAEA on India Specific Safeguards Agreement. However, in November 2007, left front gave green signal to the government to start negotiations with the Agency with this understanding that outcome of the negotiations will be placed before UPA-Left joint committee for approval as Prakash Karat, General Secretary of CPI (M) stated that, "we have come to an understanding that the government can go to the IAEA secretariat. But outcome of the talks would be brought to the committee before moving to the IAEA board of governors."[22] However, when negotiations were completed and the text of the Agreement was prepared, government did not keep its word and approached IAEA Board of Governors for its approval without consent of left front. All this made left parties to pull out of the government and government was forced to face a trust vote in Parliament which it won by a narrow margin. Expressing a deep resentment over the decision of the government, Prakash Karat termed the government's move "nothing but a shocking betrayal of moral commitment not only to the left parties, but to the country and people."[23]

India-IAEA Safeguards Agreement is based on IAEA safeguards template INFCIRC/66/Rev.2. It is an "umbrella agreement" covering multiple civil nuclear facilities and allows India to add facilities to the Agreement identified by India as civilian in future and in this context, it is an improvement over INFCIRC/66 type facility specific agreements. The Agreement which is spread over 23 pages, containing 11 sections and 130 paras/articles was the outcome of several rounds of negotiations between DAE and IAEA Secretariat.

The Agreement was made public on 9 July, 2008 and was approved by IAEA Board of Governors with consensus on 1 August 2008. The Agreement was formally signed on 2 February, 2009 by IAEA Director General Mohamed ElBaradei and Indian Ambassador Saurabh Kumar and it entered into force on 11 May, 2009. Under the Agreement, India will place under safeguards all the facilities listed in 2006 "separation plan" in a phased manner and any other facility declared by India as civilian. All these facilitates will be listed in the Annex to the Agreement. Imported nuclear material will also be placed under safeguards but it will not be listed in the Annex.

Although the Agreement was approved with consensus by the Board of Governors and no member of the Board really tried to block the Agreement, but the passage of the Agreement was not altogether a "smooth affair" as several Board members raised important queries on a number of issues pertaining to the Agreement. The most important issue which disturbed most of the members and on which clarification was sought was the reference made to "corrective measures" in the preamble of the Agreement. The countries which queried about "corrective measures" included Norway, Sweden, Ireland and Canada. A number of members also made critical statements at 1 August, 2008 meeting. For instance, Ireland stated that had there been a vote, it would have abstained. Some members threatened to adopt a tougher stand at Nuclear Supplies Group against Indian exemption from cartel's guidelines. Russia, however, stated that perpetuity principle has been maintained as the Agreement does not allow India to "withdraw facilities from safeguards at its own discretion."[24] Most of the members expressed the concern that the Agreement would set a precedent for other countries, particularly for Pakistan. Mexican Ambassador said that "one thing which worries everyone is that this could be precedent for counties to come some day before IAEA and ask for a similar agreement, including your neighbour."[25] The most serious objection to the Agreement among the Board members came from Pakistan. The objection was raised through a letter written by Pakistan's Ambassador to the IAEA, Shahbaz, to IAEA Director General on 18 July 2008. The letter blasted the Agreement on non-proliferation grounds by stating that the Agreement does not cover India's breeder and thorium programmes and against this gives her access to international fuel market. The Agreement provides for "corrective measures" which India can take if fuel supplies to it were disrupted and this violates the perpetuity principle of the IAEA safeguards system, the letter noted. Moreover, by solving India's fuel shortage problem, the Agreement provides an incentive to India to carry out nuclear tests in future and this will trigger a nuclear arms race in the subcontinent,[26] the letter further stated.

Table 3.1. List of facilities subject to safeguards under the Agreement between the Government of India and the International Atomic Energy Agency for the application of safeguards to civilian nuclear facilities.

S. No.	Facility offered for safeguards by India	Date of receipt of notification
1.	Uranium Oxide Plant (Block A),Nuclear Fuel Complex, Hyderabad	16 October 2009
2.	Ceramic Fuel Fabrication Plant (Pelletizing) (Block A), Nuclear Fuel Complex, Hyderabad	16 October 2009
3.	Ceramic Fuel Fabrication Plant (Assembly) (Block A), Nuclear FuelComplex, Hyderabad	16 October 2009
4.	Enriched Uranium Oxide Plant, NuclearFuel Complex, Hyderabad	16 October 2009
5.	Enriched Fuel Fabrication Plant,Nuclear Fuel Complex, Hyderabad	16 October 2009
6.	Gadolinia Facility, Nuclear FuelComplex, Hyderabad	16 October 2009

7. TAPS 1 – Tarapur Atomic PowerStation, Unit 1	16 October 2009
8. TAPS 2 – Tarapur Atomic PowerStation, Unit 2	16 October 2009
9. RAPS 1 – Rajasthan Atomic PowerStation, Unit 1	16 October 2009
10. RAPS 2 – Rajasthan Atomic PowerStation, Unit 2	16 October 2009
11. KK 1 – Kundankulam Nuclear PowerPlant, Unit 1	16 October 2009
12. KK 2 – Kundankulam Nuclear PowerPlant, Unit 2	16 October 2009
13. RAPS 5 – Rajasthan Atomic PowerStation, Unit 5	19 October 2009
14. RAPS 6 – Rajasthan Atomic PowerStation, Unit 6	19 October 2009
15. RAPS 3 – Rajasthan Atomic PowerStation, Unit 3	9 March 2010
16. RAPS 4 – Rajasthan Atomic PowerStation, Unit 4	9 March 2010
17. KAPS 1 - Kakrapar Atomic Power Station	3 December 2010
18. KAPS 2 - Kakrapar Atomic PowerStation	3 December 2010
19. Away from Reactor (AFR) Fuel StorageFacility, Tarapur	11 December 2012

(***Source:*** INFCIRC/754/Add.3)

Note: This is an open list and is updated when a facility is notified by India to IAEA

OPPOSITION TO THE AGREEMENT

The India-IAEA Safeguards Agreement came under a sharp criticism in both the countries. In US, non-proliferation stalwarts bitterly criticised the Agreement on non-proliferation grounds. Congressman Edward J. Markey termed the Agreement as a "shame" and said that it allows India to pull out nuclear facilities unilaterally from safeguards. He called that Agreement as pathetic and said that:

> *"[It] not only seriously undermines the Non-Proliferation Treaty, but it also sends the exact wrong message to Iran: that international nuclear safeguards are only for show. With this agreement, the IAEA has thrown its principles out of window and has abandoned its most important responsibilities... if this safeguards agreement is approved, India will be allowed to make electricity one day and bombs the next."*[27]

Daryl Kimbal of US Arms Control Association criticised the Agreement on account of its conflicting language "on what India might be able to do if it resumes testing and fuel supplies are terminated." Kimbal said that any interpretation of "corrective measures" as allowing India to withdraw its facilities from safeguards if fuel supplies to it were terminated would violate the principle of permanent safeguards. Such an interpretation is also against the Hyde Act of 2006, which states that safeguards should be permanent and in accordance with norms and practice of IAEA, Kimbal added.[28] Henery Sokolski doubted the reference made in the preamble of the Agreement to support India to build a "strategic reserve" of fuel to guard against any fuel disruption and said that it will help India to "stockpile uranium fuel against future nuclear fuel supplier cut offs that might occur...following a future test" and there is possibility of diversion of fuel from civil to military programme.[29]

In the write-up "A Non-Proliferation Disaster", Jayantha Dhanapala, former Undersecretary General for Disarmament Affairs and Daryl Kimbal underlined that the Agreement would "Undermine the nuclear safeguards system and efforts to prevent the proliferation of technologies that may be used to produce nuclear bomb material. It would also indirectly contribute to the expansion of India's nuclear arsenal with possible consequences for a nuclear arms race in Asia."[30] They further added that the Agreement would allow India to take its facilities away from safeguards if fuel supplies to it were terminated even if India breaks her moratorium on nuclear testing.[31] Similar concern was expressed by Sukla Sen of the Coalition for Nuclear Disarmament and Peace by stating that Agreement is flawed from non-proliferation standpoint as it "allows India to keep its nuclear arsenal and make fuel for nuclear weapons. It detracts from the objective of nuclear non-proliferation and disarmament and will have a negative global impact."[32]

In India, opposition against the Agreement was raised by left front and Bharatiya Janata Party (BJP). In a statement issued on 2 August, 2008, CPI (M) noted that under the Agreement, Indian nuclear facilities will be subjected to permanent safeguards and India cannot withdraw them from safeguards unilaterally. These can be removed from safeguards only if these are no longer usable for any nuclear activity.[33] CPI (M) General Secretary Prakash Karat criticized the Agreement on the grounds that it "does not take care of fuel supply, strategic reserves and corrective measures regarding fuel supply." As a regulatory body, he said IAEA has no obligation towards India either interns of fuel supply guarantees or helping it to build a strategic fuel reserve.[34]

A heavy attack on the Agreement was made through a joint statement issued by left parties on 11 July, 2008. The statement noted that the Agreement does not guarantee uninterrupted fuel supplies, strategic fuel reserve and right to take "corrective measures" in case of termination of fuel supplies to India. It is not legally binding on IAEA to provide these rights to India because reference to all these rights is made in the preamble of the Agreement and not in operative sections, the statement said. Against this, it is properly spelt out in the operational sections of the Agreement that a facility can only be withdrawn form safeguards when jointly determined by IAEA and India and if it is no longer usable for a nuclear activity,[35] the statement further said. Left also claimed that the Agreement does not treat India as a Nuclear Weapon State, because under IAEA safeguards system Nuclear Weapon States have a right to remove their facilities from safeguards but this right has been denied to India. Moreover, it is the IAEA Board of Governors which is the final arbiter as well as dispute settlement body. So, on the question of interpretation of the Agreement, the Board will have final word and any sort of noncompliance from India will make the Board to refer matter to United Nations Security Council.[36]

BJP opposed the Agreement on the grounds of national security and said that the Agreement does not recognise India as Nuclear Weapon State and would cap India's nuclear weapons programme.[37] While addressing a press conference, senior BJP leaders Yashwant Sinha and Arun Shourie criticised Prime Minister Dr. Manmohan Singh for keeping the Agreement a secret and not evolving a national consensus on the issue. They also stated that the Agreement does not guarantee uninterrupted fuel supplies to India, arms IAEA inspectors with rights to conduct special verification visits on India's nuclear facilities and there is a vague reference to the "corrective measures" in the draft Agreement.[38]

CONTENTIOUS ISSUES

The India-IAEA Safeguards Agreement, a prerequisite for the operationalisation of the India-US Civil Nuclear Cooperation Agreement became a debatable subject in India and outside. The Agreement which is concerned with safeguarding India's civil nuclear facilities came under heat because of certain contentious issues linked to the Agreement like what will be the implications of the Agreement for India's strategic nuclear programme? Does it recognize India as a Nuclear Weapon State? Does it guarantee uninterrupted fuel supplies to India? And more importantly, can India unilaterally withdraw her nuclear facilities from safeguards in future? No doubt, the Agreement is clear about first two issues that Agency will not interfere with India's nuclear weapons programme and it has no role to play regarding the recognition of India's nuclear weapon status but the ambiguity prevails regarding "fuel supply guarantees" and "termination of safeguards".

FUEL SUPPLY ASSURANCES

"Fuel supply assurances" for the life time of safeguarded nuclear reactors is an issue on which Agreement faced a lot of criticism in India. India's official stand has been that "fuel supply assurances" have been guaranteed by the Agreement. The claim is based on the reference made to this effect in the preamble of the Agreement which reads:

> *"An essential basis of India's concurrence to accept Agency safeguards under an India-specific safeguards agreement...is the conclusion of international cooperation arrangement creating the necessary conditions for India to obtain access to international fuel market, including reliable, uninterrupted and continuous access to fuel supplies from companies in several nations, as well as support for an Indian effort to develop a strategic reserve of nuclear fuel to guard against any disruption of supply over life time of Indian reactors."*[39]

However, this claim has been challenged on the grounds that preambular references are not legally binding, as preamble is mainly used to provide background and context to the agreement. It cannot create any additional rights or obligations which do not find any mention in the operative sections of the Agreement.[40] This was confirmed by Mohamed ElBaradei in his August 1, 2008 statement that the Agreement's preamble provides for contextual background. [41] Refuting the government claims, Brahma Chellaney, a strategic expert argues that:

> *"It is well established in international law that no rights flow from preambular reference which are usually intended to provide background or context to an agreement—unless such reference is spelled out in the body of the accord or treaty... mere appearances in the preamble of the terms "reliable, uninterrupted and continuous fuel supplies", "a strategic reserve of nuclear fuel"... cannot be used to claim rights or obligations not contained in the operative parts of the agreement".*[42]

A best example of this is provided by NPT in which the reference to achieve the goal of nuclear disarmament is made in the preamble of the treaty and not in the operative parts, that is why the goal of nuclear disarmament has not been achieved yet.[43]

It must also be noted that the statute of the Agency does not assign it any function regarding the supply of fuel to the states parties to any safeguards agreement. It is a regulatory body entrusted with the function to prevent nuclear proliferation by safeguarding nuclear material and activates. The Agency has no control over nuclear fuel supplies, therefore, it cannot guarantee uninterrupted supply of fuel. The issue of uninterrupted fuel supplies is to be settled down between the buyer and the seller. So, India has to negotiate fuel supply arrangements with supplier countries before placing its facilities under safeguards. This was acknowledged by Dr. B. Grover who was a part of safeguards negotiating team in a press conference on July 12, 2008 that "the Agency itself did not supply fuel. Through this particular safeguards agreement, you have to create those conditions which help us [India] to proceed further in importing fuel".[44] Dr. Anil Kakodkar has also reiterated that any fuel supply agreement "has to be between supplier and us [India]. So, we will have to build in, as we have been doing always, a strong commitment on the part of suppliers to continue, whether it is fuel or spares or whatever."[45]

TERMINATION OF SAFEGUARDS

The most contentious issue associated with India-IAEA Safeguards Agreement is the unilateral termination of safeguards. This issue emanates from the preamble of the Agreement which states that "India may take corrective measures to ensure uninterrupted operation of civilian nuclear reactors in the event of disruption foreign fuel supplies."[46] The term "corrective measures" which also exists in the text of 123 Agreement and in March 2006 "separation plan" is nowhere defined, but it has been taken to mean India's right to withdraw civil nuclear facilities form safeguards in case of disruption of fuel supplies. Dr. Anil Kakodkar defined "corrective measures" as "unspecified sovereign rights", which India will take, if fuel supplies were terminated for safeguarded reactors. When Dr. B. Grover was asked can India withdraw its reactors from safeguards, he replied that if fuel supplies were cut off, then first India under Article 52 (c) of the Agreement can "raise this is a material violation of the agreement and that itself might act as deterrent. If still a danger of disruption arises, we have here [in the Agreement] the combination [Articles] 29, 30f, 10, 4 and the preamble, this will help us with whatever steps we want to take."[47] However, Vilmos Cserveny, the Chief negotiator for Safeguards Agreement, while briefing the Board members on July 25, 2008, made it clear that the term "corrective measures" as mentioned in the preamble of the Agreement should not be "confused with termination of safeguards", the termination provision is governed by Articles 29 and 32 of the Agreement.[48] Mohamed ElBaradei also expressed a similar argument in his introductory statement to the Board of Governors on 1 August 2008, when he said that:

> *"As with other safeguards agreements between the Agency and Member States, the (Indian) agreement is of indefinite duration. There are no conditions for discontinuation of safeguards other than those provided by safeguards agreement. The termination provisions in the agreement are same as for other 66-type agreements."*[49]

He further added that the specific termination clauses of the Agreement override the general clauses of the Agreement.[50]

In ElBaradei's statement two points are noteworthy. First, what INFCIRC/66 Rev.2 on which Indian Safeguards Agreement is modeled says on termination of safeguards and second, what are the "specific termination clauses" of the India-IAEA Safeguards Agreement. INFCIRC/66.Rev.2 is silent on the termination of safeguards on nuclear facilities and so far as the termination of safeguards on nuclear material is concerned the document lists six conditions in Article 26 under which safeguards on nuclear material can be terminated. The duration of safeguards on nuclear facilities is fixed by supplier and the recipient state in a bilateral agreement under which such facilities are supplied.[51] So far as "specific termination clauses" in the Agreement are concerned, here Article 29 and 32 are noteworthy. Article 29 states that "the termination of safeguards on items subject to this Agreement shall be implemented taking in to account the provisions of Gov/1621(20 August 1973)"[52] and Article 32 of the Agreement states that safeguards on a nuclear facility can be terminated when it is jointly determined by India and the IAEA that such a facility is irrelevant for safeguards. Safeguards on non-nuclear material, equipment and components can be terminated "as and when the non-nuclear material, equipment or components have been returned to the supplier or arrangements have been made by the Agency to safeguard the non-nuclear material, equipment or components in the state to which it is being transferred, or when India and the Agency have jointly determined that non-nuclear material, equipment or component in question has been consumed , is no longer usable for any nuclear activity relevant form the point of view of safeguards or has become practically irrevocable."[53]

Article 32 clearly lays down the conditions under which safeguards on a nuclear facility and nuclear material subject to the Agreement can be terminated, but Article 29 links the termination of the safeguards to GOV/1621. GOV/1621 is a restricted IAEA document which was introduced in 1973 to have greater degree of standardization with respect to duration and termination of 66 type Agreements. In this respect, the document notes that the safeguards agreements concluded between a state and the Agency must reflect two important concepts. First, the safeguards on equipment, material and facilities supplied to a state, should continue to apply till the period of "actual use of items in the recipient state" and second, the provisions regarding the termination of safeguards on supplied nuclear material, equipment and facilities should be formulated in such a way that the rights and obligations of the parties should continue to apply till safeguards on such items are terminated by the Agency.[54] In the Annex to the document, it is further stated that the rights and obligations of parties as stipulated in the agreement with respect to supplied material and items or in connection with material and items included in the inventory would continue to apply even after the termination of the safeguards, until these are removed from the inventory. However, it further states that these items could be removed from inventory and thus from safeguards only when they are consumed or are no longer usable for any nuclear activity relevant from the safeguards point of view.[55]

So, the document makes it clear that whatever comes from outside will remain under perpetual safeguards and there is no disagreement on this issue among experts. But what about indigenous facilities which India will place under safeguards. Can India remove them from safeguards unilaterally is a contentious issue. Some experts and analysts believe that India can remove its indigenous facilities from safeguards if all imported material subject to safeguards is removed from these facilities.[56] Holding the same argument Siddharth Varadarajan argues that with respect to termination of safeguards on indigenous reactors, it is paragraph 3 of the GOV/1621 which is relevant. It states that with respect to equipment, facilities and no-nuclear material the conditions of removal could be based on paragraph 26 of INFCIRC/66/Rev.2 and according to Varadarajan it is paragraph 26(f) of INFCIRC/66/Rev.2 which India can invoke to withdraw its indigenous facilities from safeguards if a situation demands. It states that:

> *"Safeguards on nuclear material shall no longer apply if the conditions specified in the safeguards agreement, pursuant to which it was subject to Agency safeguards no longer apply, by the expiration of the agreement or otherwise."*[57]

When read with paragraph 3 of GOV/1621 it makes it clear that an indigenous facility can be removed from safeguards when condition pursuant to which it was subject to safeguards ceases to apply. One of the conditions as mentioned in the preamble of the India-IAEA Safeguards Agreement under which India agreed to place its indigenous reactors under safeguards is permanent supply of fuel and if fuel supply is terminated India can claim that such a condition no longer applies, so there is no justification for keeping its indigenous reactors under safeguards.[58]

However, this view has been contested by many analysts.[59] The central thread running through their argument is that India cannot withdraw any of its facilities including indigenous ones form safeguards as Article 11 and 32 of the Agreement provide for perpetuity of safeguards even in the case of indigenous reactors. Prabir Purkayastha, a noted analyst on disarmament and energy issues holds that the termination of safeguards on indigenous facilities is governed by Articles 11(a) and 32 of the Agreement. Article 11(a) of the Agreement provides that the items subject to the Agreement shall be "Any facility listed in the Annex to this Agreement" and Article 32 states that:

> *"Safeguards shall be terminated on a facility listed in the Annex after India and the Agency have jointly determined that the facility is no longer usable for any nuclear activity relevant from the point of view of safeguards".*[60]

So, once a facility is listed in the Annex to the Agreement whether imported or indigenous one it can be removed from safeguards only if it is jointly agreed by the Agency and India that such a facility is no longer usable for any nuclear activity and there is no question of India unilaterally withdrawing any facility from safeguards.[61]

Lets come back to Article 29 of the Agreement which states that the termination of safeguards on items subject to this Agreement shall be implemented taking into account the provisions of GOV/1621. The word "items" used in Article 29 includes as per Article 11(a) of the Agreement any facility listed in the Annex to the Agreement. So when Article 29 is read with Article 11(a) of the Agreement, it implies that once a facility —imported or indigenous— is listed in the Annex it will remain under permanent safeguards as guaranteed by GOV/1621, which otherwise covers only supplied material and facilities.

While going through different provisions of the Agreement it becomes clear that the imported reactors and material will stay under safeguards permanently and as such there exists no disagreement on this issue among the analysts. However, regarding the indigenous facilities which for the first time have been offered for safeguards, disagreement prevails and the ambiguous language of the Agreement puts the reader in a fix. If a situation arises in future when India will be compelled to withdraw its indigenous reactors from safeguards, which is only possible if fuel supplies to it are terminated, then much more will depend on the interpretation of the Agreement. As per the statements of IAEA officials, it is explicitly clear that IAEA will not compromise with the "perpetuity principle". Since India has accepted IAEA as final arbiter, so there will be limited room for India to manoeuvre if interpretation of the Agreement comes into play and any non-compliance from India can make IAEA to take appropriate measures to ensure perpetuity of safeguards, including the referral of the matter to United Nations Security Council which is the sole authority to implement the safeguards.

THE ADDITIONAL PROTOCOL

Another watershed event towards the operationalisation of India-US Civil Nuclear Cooperation

Agreement was the signing of Additional Protocol by India, thereby fulfilling all her commitments made in the Indo-US joint statement of 18 July, 2005. Besides the joint statement of 2005, India's obligation to conclude an Additional Protocol with Vienna based nuclear watchdog, IAEA also finds a mention in the US legislation, Hyde Act which states that "India must make a substantial progress toward concluding an additional protocol consistent with IAEA principles, practices and policies that would apply to India's civil nuclear programme before the United States could engage in nuclear trade with India."[62] The text of the Protocol was finalized in mid-February 2009 after five rounds of negotiations between IAEA Secretariat and India's Department of Atomic Energy (DAE). The Protocol was approved by IAEA Board on 3 March 2009 and it was signed by India on 15 May 2009. The Protocol will come into force when ratified by Indian Parliament. Currently there are ninety countries which have such agreements in force including five nuclear weapon states. It may be noted that the signing of an Additional Protocol is not mandatory for a country; rather it is a voluntary affair.

The exact picture of India's Additional Protocol will come into light only when it is ratified by Indian Parliament, however, from different sources it has come to light that the Additional Protocol signed by India is a clear departure from Model Additional Protocol (INFCIRC 540), which is viewed by international community as a significant tool to strengthen control over the proliferation of nuclear material and activities in a number of significant areas. Under the Protocol as signed by India, it is not bound to provide any sort of information to IAEA regarding its unsafeguarded nuclear programme. Para 1(b) of the agreement clearly states that the "[P]rotocol shall be implemented in such a way that it does not hamper, hinder or interfere with any activities involving the use of non-safeguarded material and equipment."[63] Inter alia, the Protocol does not find any mention of sharing the information on nuclear fuel cycle related to research and development including uranium mining, fuel fabrication, enrichment plants and nuclear waste sites with IAEA. The Protocol also does not contain any "complementary access" provisions under which IAEA can carry out short notice inspections and take environmental samples to reveal undeclared nuclear activities. All these provisions form an important part of standard template (INFCIRC 540) for Additional Protocols. Under the Additional Protocol, India is bound to provide IAEA inspectors multiple-entry visas and to allow the Agency free "communications... including attended and unattended transmission of information generated by the Agency containment and/or surveillance or measurement devices which will already be put in place in India's safeguarded facilities." This means that IAEA will get the information generated by its devices placed in safeguarded facilities rather than physically accessing that information during a visit to a safeguarded facility.[64]

India's Additional Protocol is also less burdensome when compared with the Additional Protocols of five nuclear powers. The Additional Protocols signed by five nuclear powers place them under voluminous reporting obligations, but India on the other hand is obliged to provide the Agency information regarding its nuclear exports only.[65]

NUCLEAR SUPPLIERS GROUP (NSG): A BRIEF INTRODUCTION

Nuclear Suppliers Group is a multinational body which was formed in response to India's nuclear test of 1974 to control the export of nuclear material and technology that could be used for the development of nuclear weapons. The formation of the group was proposed by the United States with the objective to strengthen control over nuclear material and equipment to prevent their diversion in recipient states and towards the fulfillment of this objective NSG has developed two sets of guidelines. Part I guidelines were adopted in 1978 after a series of meetings by the group and were published as INFCIRC/254. These guidelines cover those items and material that could be directly used to produce nuclear weapons. These are called "trigger list" items. Part II guidelines were adopted in 1992 in response to Iraq's clandestine nuclear weapons programme that used the items that were not covered by Part I guidelines. These guidelines were published as INFCIRC 254/ Rev.1/ part II. These guidelines pertain to dual-use items i.e. those items that have both nuclear and non-nuclear applications. The first set of guidelines regulate the export of items that include (i) nuclear material (ii) nuclear reactors and equipment (iii) non-nuclear material for reactors (iv) plant and equipment for reprocessing, enrichment, fuel fabrication and heavy water production, and (v) technology related to above mentioned items. These guidelines instruct a supplier state that it should authorize the transfer of "trigger list" only when it receives a formal assurance from a recipient state that such items would not be used in any explosive device (Art.2) and such items should be placed under effective physical protection to prevent their theft and illegal transfer (Art. 3(a)). In the case of Non-Nuclear Weapon States such items should only be transferred when a recipient state has a full scope safeguards agreement in force. An important feature of these guidelines is the "non-proliferation principle" which states that:

> *"Notwithstanding other provisions of the guidelines, suppliers should authorize*

transfer of items or related technology identified in the trigger list only when they are satisfied that the transfers would not contribute to the proliferation of nuclear weapons or other explosive devices or be diverted to acts of nuclear terrorism."[66]

The second set of guidelines that regulate the supply of dual use goods stipulate that a supplier country should not authorize the transfer of such goods if they are

- "for use in non-nuclear weapon state in a nuclear explosive activity or an unsafeguarded nuclear fuel cycle activity" or
- "in general when there is an unacceptable risk of diversion to such an activity, or when transfers are contrary to the objective of averting the proliferation of nuclear weapons" or
- "when there is an unacceptable risk of diversion to acts of nuclear terrorism" (Art. 2)[67]

Moreover, Art. 4 of the guidelines states that before authorizing the transfer of dual use goods suppliers should exercise "prudence" and should take in to account relevant factors including:

(i) Whether the recipient state is signatory to Nuclear Non-Proliferation Treaty or other regional legally binding non-proliferation agreement and has an IAEA Safeguards Agreement in place covering all the peaceful nuclear activities of that state.

(ii) Whether the recipient state that is not signatory to NPT or any other regional non-proliferation arrangement has any nuclear facility either operational/being designed or constructed that is not or will not be covered by Agency's safeguards mechanism.[68]

Presently, NSG consists of 47 members and any country that supplies nuclear material, equipment and technology to other countries can apply for membership. But in order to become a member of NSG, a country should have an impeccable non-proliferation record and a number of factors are taken into account before granting membership to a country; these include:

- The ability of a country to export items covered by NSG guidelines.
- Adherence to NSG guidelines.
- Compliance to non-proliferation obligations as laid down in NPT or other regional non-proliferation instruments.
- Support to the international efforts to prevent proliferation of weapons of mass destruction.[69]

NSG takes all its decisions by consensus. So, any decision whether related to the amendment of guidelines or inclusion of new members must be agreed by all the members.

NUCLEAR DEAL AND ITS CLEARANCE BY NSG

Technically, India is a Non-Nuclear Weapon State (NNWS) and it has not signed a Comprehensive Safeguards Agreement (CSA) with IAEA, therefore, it is not eligible to trade with other countries of the world in civil nuclear field. Guidelines of the NSG, cartel that regulates nuclear trade make it mandatory for a Non-Nuclear Weapon State to sign a Comprehensive Safeguards Agreement with IAEA to engage in civil nuclear trade with other countries of the world. So, in order to exempt India form the requirement of Comprehensive Safeguards Agreement and to fulfill the commitment recorded by President George W. Bush in July 2005 joint statement that "US will work with friends and allies to adjust international regimes [a reference made to NSG] to enable full civil nuclear energy cooperation and trade with India...", US approached NSG with a resolution seeking adjustment of cartel's guidelines for India. The initial draft of the waiver was placed before NSG plenary on 21-22 August 2008; but the proposal was shot down by the group on the non-proliferation grounds. At the meeting, a significant number of countries proposed amendments to the draft of the waiver and demanded that strict conditions must be imposed on India before granting it any exemption. Following the rejection of the first draft India and US prepared a revised version of the first draft which was placed before NSG on 4 September 2008. However, draft faced a stiff resistance from six like minded countries— Austria, the Netherlands, Ireland, Switzerland, Norway and New Zealand — on the grounds that the draft of the waiver did not reflect their concerns voiced at 21-22 August plenary meeting. These countries demanded that in the draft waiver, three conditions must be incorporated:

(i) Termination of nuclear trade with India if India tested a nuclear device.

(ii) A periodic review of India's compliance to the non-proliferation commitments.

(iii) Non transfer of sensitive nuclear technologies to India.[70]

In order to allay the concerns of these countries, then Indian External Affairs Minister issued a statement on September 5, 2008 reiterating India's non-proliferation commitments. The statement proved as a turning point and provided necessary "basis" for reaching a consensus at the meeting. To quote a European diplomat: "If you ask me to identify a turning point in our work, it was the circulation of that statement."[71] The text of the waiver makes an explicit reference to the statement in paragraph 3. The statement reads:

> *"We [India) remain committed to a voluntary, unilateral moratorium on nuclear testing. We do not subscribe to any arms race, including a nuclear arms race. We have always tempered the exercise of our strategic autonomy with a sense of global responsibility. We affirm our policy of no-first-use of nuclear weapons. We are committed to work with others towards the conclusion of multilateral Fissile Material Cut-off Treaty in the Conference on Disarmament that is universal, non-discriminatory and verifiable."*[72]

Besides the statement of Indian External Affairs Minister it was the American diplomatic might which played an outstanding role in granting waiver to India. United States used its every diplomatic weapon to persuade the hold-out countries to grant India a specific exemption. To quote David Mulford, former US Ambassador to India: "It was the biggest diplomatic effort we have witnessed since 1980s."[73] The waiver finally granted to India after three days of roller-coaster deliberations exempts India from certain requirements of part I and part II of the guidelines respectively. Paragraph 3(a) of the waiver notes:

> *"Notwithstanding paragraphs 4(a), 4(b) and 4(c) of INFCIRC 254/Rev.9/part I, participating Governments may transfer trigger list items and/or related technology to India for peaceful purposes and for use in IAEA safeguarded civil nuclear facilities, provided that the transfer satisfies all other provisions of INFCIRC 254/ part I, as revised..."*[74]

Likewise paragraph 3(b) of the waiver states:

> *"Notwithstanding paragraphs 4(a) and 4(b) of INFCIRC/254/Rev.7/part 2, participating Governments may transfer nuclear related dual-use equipment, materials and software and related technology to India for peaceful purposes..."*[75]

The waiver which lifted a 16 year old nuclear trade embargo on India evoked a mixed response. While it was hailed by Indian officials but faced a sharp criticism from certain Indian political establishments, non-proliferation stalwarts and strategic experts. Indian Prime Minister Dr. Manmohan Singh Manmohan Singh called the waiver a "forward-looking and a momentous decisions" and said that the waiver "marks an end of India's decades long isolation from nuclear mainstream and of technology denial regime."[76] Indian Atomic Energy Negotiator, R.B. Grover called it a clean waiver and said "Changes made in the draft is mutually agreed upon. We have no problem with the draft."[77] External Affairs Minister, Pranab Mukherjee asserted that the waiver meets all the Indian expectations and is "in accordance with the commitments given to the Parliament and people of India and is consistent with India's national interests."[78] Atomic Energy Commission Chairman Anil Kakodkar termed the waiver a "breakthrough" and said that it is an "acceptable arrangement which meets our requirements." He further added that the waiver imposes no new conditions on India and is consistent with what India was looking for.[79]

However, the waiver was bitterly criticized by Indian Political parties particularly by BJP and left wing parties and US non-proliferationists. In a statement CPI (M) said that the waiver is neither "clean nor unconditional" and has converted India's unilateral voluntary moratorium on nuclear testing into a multilateral commitment. Party argued that the waiver is in full conformity to the provisions of the Hyde Act and by accepting it India has become a party to US efforts to deny the rights of full nuclear fuel cycle to other countries – a reference made to Iran.[80] In a brief statement, General Secretary of the party Prakash Karat said that the waiver "is surrender towards operationalsing the nuclear deal." Expressing a deep sense of dissatisfaction over the waiver, Communist Party of India (CPI) termed the waiver a "black day" for India and said that the waiver will prove detrimental for India's indigenous nuclear programme. [81] BJP complained that waiver has pushed India into "non-proliferation trap" and India has surrendered her right to test a nuclear device in future. Party leader Yashwant Sinha described September 6, the day on which waiver was granted to India "a day of historic shame for India" and said that the government has surrendered country's national interests.[82]

The non-proliferation activists protested against the waiver on the non-proliferation grounds. In a letter written to German Foreign Minister Frank Walter, non-proliferation activists called the waiver a "non-proliferation disaster" and urged that in the draft waiver more stringent conditions must be imposed on India to prevent the erosion of NPT regime.[83] Democratic Congressman Edward Markey said that the waiver "'set(s) a dangerous precedent' and will certainly damage future non-proliferation efforts".[84] Ellen Tauscher, member of House of Representatives, claimed that the waiver is dangerous in the sense that it gives India rights to engage in nuclear trade with other countries of the world without placing responsibilities on it.[85] Stimson Center's Michael Krepon argued that the waiver is silent on the issue of "nuclear testing" and in future it will be difficult for international community to punish India if it violated non-proliferation norms.[86] Daryl Kimbal of Arms Control Association said that the waiver has undermined the efforts to contain the nuclear programmes of rogue states and it will make it impossible to "win the much needed support of other countries of the world to strengthen the non-proliferation regime."[87]

NSG WAIVER: IS IT CLEAN AND UNCONDITIONAL?

Before and during the NSG deliberations a consistent demand of Indian interlocutors was that the India-specific waiver should be "clean and unconditional". "Clean" in the sense that it should provide for full civil nuclear cooperation and "unconditional" in the sense that it should not impose new conditions on India beyond those as stipulated in cartel's guidelines.[88] So, the waiver granted to India by NSG and the "conditions" on which it was granted raise a fundamental question: Is it "clean and unconditional" as sought by New Delhi? The answer probably is "no". A proper and careful reading of the waiver and NSG deliberations make it amply clear that the waiver is neither "clean" nor "unconditional".

In the first place, waiver has converted India's unilateral non-proliferation commitments particularly her unilateral voluntary moratorium on nuclear testing into multilateral commitments. These commitments were reiterated by Indian External Affairs Minister in his September 5 statement and NSG made these commitments a "basis" for exempting India by making an explicit reference to them in paragraph 3 of the waiver. Any departure form these commitments could have serious repercussions for India as it was stressed by a number of countries in their statements, issued after the waiver was adopted. For instance New Zealand Prime Minister, Helen Clark stated that:

> *"It was India's commitment to unilateral moratorium of nuclear testing that convinced New Zealand to back the nuclear Suppliers Group Waiver to New Delhi... Any resumption of testing could trigger a meeting and cause a decision to review."*[89]

Japan stated that the waiver was granted to India on the condition that India would continue her moratorium on nuclear testing and cautioned that any violation of this commitment will lead to the termination of nuclear cooperation with India.[90] This point was also stressed by a number of countries in their statements. In the event of India violating its non-proliferation commitments supplier courtiers can invoke paragraph 3(e) of the waiver which provides for holding consultations if circumstances demand and act "in accordance with paragraph 16 of the guidelines."[91] Paragraph 16(c) of the guidelines provides for holding consultations if one or more supplier countries believe that "there has been a violation of supplier/ recipient understanding particularly in the case of explosion of nuclear device" by a recipient state and calls upon the suppliers to take appropriate action including termination of nuclear supplies.[92]

Secondly, the waiver does not provide for full civil nuclear cooperation with India. It links the export of sensitive nuclear technologies to paragraph 6 and 7 of NSG guidelines. The provisions of paragraph 6 of INFCIRC /254/ REV.10/ p I make it amply clear that to receive sensitive nuclear technologies is a prerogative of those countries that are parties to NPT and have an impeccable non-proliferation record. Moreover, there was a general understanding among the participating governments at the plenary meeting not to supply such technology to India. A report published in the *Washington Post* of September 12, 2008 reveals that during the deliberations member countries privately agreed on non-transfer of enrichment and reprocessing technology to India and it was this understanding which made hold-out countries to reverse their stand.[93] Moreover, at the plenary meeting a number of countries issued national statements reiterating their pledge not to transfer enrichment and reprocessing technology to India.[94] In the NSG, US is a dominant player and it is the policy of US not to transfer such technology to other countries and in the case of India, US legislation, Hyde Act instructs it to work with other NSG countries to restrict the transfer of sensitive technology to India (section 105(a) (5).[95] So, every US government will ensure that such technology is not transferred to India by any supplier.

Thirdly, India has agreed to abide by the NSG guidelines. This is a serious commitment because India is obliged to follow not only the present guidelines but also the amendments made in the guidelines in future without having a say in the adoption of such amendments because India is not a member of the group. So, in future if cartel imposes new conditions on India for carrying out nuclear trade with NSG countries, India will be left with no choice but to abide by them.

Lastly, there will be a periodic review of India's compliance to non-proliferation commitments. Paragraph 4(c) of the waiver calls up on participating governments to consult each other on regular basis "for the purpose of considering matters connected with the implementation of all aspects of this statement taking in to account relevant international commitments on bilateral agreements with India" and a non-compliance on India's part can force the cartel to review its September 6 decision.

WAIVER AND THE NEW NSG GUIDELINES

On 24 June 2011, Nuclear Suppliers Group in its plenary meeting held at Dutch town of Noordwijk, Netherlands adopted new guidelines pertaining to transfer of enrichment and reprocessing (ENR) technology. The new guidelines were in the pipeline from a long time but disagreement among the members of the group eluded consensus on the issue for over ten years. The idea of adopting stringent norms for ENR transfers was first floated by US in 2001 in response to supply of fifty-seven tons of nuclear fuel by Russia to India by invoking the "safety clause" of NSG guidelines. Realizing that such transfers could also take place in future involving the transfer of sensitive technologies, US proposed amendments in the guidelines pertaining to supply of ENR equipment and technology.[96] September 11 attacks on US and unmasking of A.Q. Khan network which was allegedly involved in the proliferation of sensitive nuclear technology further spurred the need to tighten the norms pertaining to the export of enrichment and reprocessing technology. In a speech delivered in 2004, US President George W. Bush called upon the members of the NSG that:

> *"[They] should refuse to sell enrichment and reprocessing equipment and technology to any state that does not already possess full scale, functioning enrichment and reprocessing plants. This step will prevent new states from developing the means to produce fissile material for nuclear bombs. Proliferators must not be allowed to cynically manipulate the NPT to acquire the material and infrastructure necessary for manufacturing illegal weapons."*[97]

But US failed in its bid to obtain necessary support from NSG to restrict the supply of ENR technology along the lines envisaged by George W. Bush. However in 2008, due to US diplomatic push NSG in its Consultative Group meeting succeeded in framing the draft of new guidelines called "clean text".[98] Though "clean text" could not generate necessary consensus among the member countries but provided necessary basis for further negotiations on the issue and finally in 2011 group succeeded in generating consensus on new guidelines. The new guidelines are by and large same as "clean text" with some minor modifications.

The new guidelines were adopted by amending paragraph 6 and 7 of INFCIRC /254/ REV.9/part I. Prior to the amendment paragraph 6 of the guidelines stated that:

> *"Suppliers should exercise restraint in the transfer of sensitive facilities, technology and material usable for nuclear weapons or other nuclear explosive devices. If enrichment or reprocessing facilities, equipment or technology are to be transferred suppliers should encourage recipients to accept an alternative to national plants, suppliers involvement and/or other appropriate multinational participation in resulting facilities."*[99]

Paragraph 7 stated that in the case of transfer of enrichment technology, the recipient state must not enrich uranium beyond twenty per cent without the consent of supplier.[100] Under the revised guidelines, a recipient state has to fulfill a number of conditions to receive enrichment and reprocessing technology. These conditions as laid down in paragraph 6 of INFCIRC 254/ Rev.10/ P I include:

- A state must be party to NPT and it must be in full compliance to the obligations of the treaty.

- It must not have been cited by IAEA as being in breach of its Safeguards Agreement.
- It must be adhering to NSG guidelines and must be implementing effective export controls as identified in UNSC Resolution 1540.
- It must have in place Comprehensive Safeguards Agreement and an Additional Protocol or pending this regional arrangement approved by IAEA.
- It must be in adherence to international nuclear safety norms.
- It must have concluded with a supplier state an intergovernmental agreement covering non-explosive uses, effective safeguards in perpetuity and retransfer.[101]

In paragraph 7, twenty per cent enrichment criteria has been retained but some additional conditions have been incorporated.

The revised guidelines/rules have become a source of great concern for New Delhi as they limit the supply of enrichment and reprocessing technology to states that are parties to NPT and India does not qualify this criterion. The new guidelines have been viewed by India as a "rollback" of the waiver won by it in September 2008 and being contrary to Indo-US Civil Nuclear Deal concluded in 2008. Providing a sharp response to NSG decision, Indian External Affairs Minister S.M. Krishna told Parliament that NSG waiver granted to India in 2008 form the "basis and overarching framework that governs the cooperation in civil nuclear matters between India and NSG. The issue is the full implementation of that understanding. This is what we expect and our major partners are committed to."[102] Former Chairman of India's Atomic Energy Commission called the decision as "betrayal of the exemption NSG had granted to India." He further said that "In the bilateral 2008 Indo-US civil nuclear deal, there was some forward looking language, the understanding was that even if it is not possible now [transfer of ENR technology], it would be possible in future. The new NSG guidelines are completely contrary to that spirit."[103] To allay Indian concerns, three main supplier countries—US, Russia and France— have assured India that the new guidelines will not affect 2008 Indian exemption. The assurance has come through the statements issued by these countries after the agreement on revised guidelines was reached. However the statements issued be these countries are ambiguous and it seems unlikely that they will any way help India to bypass the new rules. Consider the US statement first. It states:

> *"The NSG's NPT reference including those in the ENR guidelines in no way detract from the exemption granted to India by NSG in 2008 and in no way reflect upon India's non-proliferation record. These references reinforce the commitment of NSG members to prevent transfer of items that could be used for weapon purposes ...."*[104]

The other part of the statement underlines that the "Efforts in the NSG to strengthen controls on the transfer of ENR are consistent with longstanding US policy that predates Civil Nuclear Agreement [with India]..."[105] The longstanding US policy on the transfer of ENR technology is that it does not supply such technology to any state and in the case of India, officials of the State Department have time and again reiterated that civil nuclear cooperation with India will not include transfer of ENR equipment and technology. In 2005, Under Secretary Robert Joseph testified before the Senate Foreign Relations Committee that: "We do not intend to provide enrichment and reprocessing technology to India. As the

President said in February 2004, enrichment and reprocessing are not necessary for nations seeking to harness nuclear energy for peaceful purposes. We do not currently provide enrichment and reprocessing equipment to any country."[106] This was also confirmed by US Secretary of State in April, 2006 when she informed Senator Richard Lugar that US will not export ENR technology to India.

Consider the statement of Russian Foreign Minister:

> *"The decision to strengthen sensitive nuclear export control adopted at NSG plenary session in 2011 does not affect in any way the September 2008 decision of the group to unfreeze peaceful nuclear cooperation with India."*[107]

It must be noted that Russia has been adhering to new NSG guidelines even prior to their adoption in 2011. In compliance to the statement issued by G8 countries in 2009 at L'Aquila, Italy which declared that G-8 countries would "implement" as a national policy the "Clean text" on the transfer of ENR technology, [108] Russian government adopted a government decree No.992 in December 4, 2009 that allowed supply of ENR technology to states that are signatories to NPT. The decree was valid for one year but was renewed in 2010.[109] Though Russia has assured India that new rules will not affect 2008 exemption, but at the same time it is insisting that any nuclear trade between two countries will be within the parameters of its national laws and international obligations. Prior to the NSG decision of 2011, Russia was looking to set up a reprocessing plant on Indian soil in collaboration with India as part of negotiations going on between two sides since 2010 on the transfer of ENR technology to India. But after 2011 agreement, a significant change came in Russia position, Russia now insisting to shift the project to Russia which is a clear indication that it does not want to annoy its NSG partners. To quote Russia Ambassador to India, Alexander Kadakin, "Russia has offered to host an ENR unit on its soil and offered shares to India, this formula will enable Russia to comply with its obligation to Indian friends and not violate international obligations...."[110] The statement issued by French Ambassador to India, Jerome Bonnafont is somewhat similar to Russian statement. It states that: "France confirms that this NSG decision no way undermines the parameters of our bilateral cooperation and is committed to full implementation of cooperation agreement on the development of peaceful uses of nuclear energy signed on September 30, 2008."[111] The parameters of bilateral cooperation between two sides do not involve ENR transfers and nuclear cooperation agreement of September 30, 2008 does not cover such transfers either and there is no indication that such transfers will be made in future. Prior to June 24, 2011 decision French officials made it clear that France will not supply ENR equipment and technology to India and it must be stressed that "NPT membership criteria"; the first condition of new guidelines was the French proposal.[112]

CONCLUSION

The conclusion of India-IAEA Safeguards Agreement and adjustment of NSG guidelines, two important milestones covered towards the operationalisation of India-US Civil Nuclear Cooperation Agreement have brought to light certain important facts having serious implications for India. In the first place, India-IAEA Safeguards Agreement reveals that the "uninterrupted fuel supplies" is an issue which India has to settle unilaterally with a supplier country and IAEA is not liable for providing such guarantees to India as it is the Agency which is primarily concerned with preventing the diversion of nuclear material and equipment and has no control on fuel supplies. Secondly, once the nuclear facilities, whether imported or indigenous are placed under safeguards they will remain under safeguards in perpetuity and India cannot unilaterally withdraw them from safeguards. Though some ambiguity prevails regarding indigenous reactors but the way IAEA officials have interpreted the Agreement, there seems little scope that India can remove them from safeguards either. Thirdly, the waiver granted to India by NSG is conditional. It depends upon India's compliance to its non-proliferation commitments and any violation of these commitments can make the cartel to rollback the waiver. Lastly, by tightening the control on the export of sensitive enrichment and reprocessing technology by limiting its access to states that are parties to NPT, NSG has put brakes on full civil nuclear cooperation with India.

REFERENCES

1 Bertrand Goldschmidt, "The Origins of International Atomic Energy Agency," pp. 12-13. Available at http//www.iaea.org/publications/Magzines/Buletein/1940128121, (accessed on 1 August 2012).

2 "Atoms for Peace Speech," (1953), Address by Mr. Dwight D. Eisenhower, President of United States of America to the 47th Plenary Meeting of the United Nations General Assembly. Available at http//www.iaea.org/about/ atoms for peace _speech. htm, (accessed on 10 October 20120).

3 "The Statue of the IAEA." Available at: http//www.iaea.org/About/Statue.htm, (accessed on 7 August 2012).

4 *Ibid.*

5 "International Atomic Energy Agency." Available at http//en.wiki.org.wiki/international_Atomic_Energy_Agency, (accessed on 3 August 2012).

6 *Ibid.*

7 Gabriel Eugenio Andres, May-December, 2008, "The

International Atomic Energy Agency's Safeguards System," *International Law Review*, p. 94.

8 Ben Sanders, 2004, "IAEA Safeguards and the NPT," p. 44. Available at http://www. unider.org/pdf/articles/pdf/art2189.pdf, (accessed on 9 December 2012).

9 Jack Boureston and Charles D. Ferguson, "Strengthening Nuclear Safeguards: Special Committee to Rescue?." Available at www.armscontrol.org/print/951, accessed on (1 September 2012).

10 Gabriel Eugenio Andres, *op. cit.*, pp. 96-97.

11 Kalpana Chittaranjan, "The IAEA as a Safeguards Regime." Available at http://www.nuclearfiles.org/menu/key-issues/nuclear-energy/history/article—chitranjan-kalpana.htm, (accessed on 22 December 2012).

12 Pierre Goldschmidt, 1995, "The IAEA Safeguards System Moves into 21st Century," *IAEA Bulletin* 14/4, pp. 19-20

13 "93" means the decision to strengthen safeguards was taken in 1993 and "+2" means it was to be completed in 1995.

14 Kalpana Chittaranjan, *op. cit*.

15 "IAEA Safeguards: Stemming the Spread of Nuclear Weapons." Available at http://www.iaea.org/publications/Factsheets/English/S1safeguard.pdf, (accessed on 12 August 2012).

16 "The Safeguards System of International Atomic Energy Agency." p. 2. Available at http://www.iaea.org/ourwork/sv/safeguards/documents/safeguardssystem.pdf, (accessed on 2 June 2012).

17 Jan Priest, 1995, "IAEA Safeguards and NPT: Examining interconnections,", *IAEA Bulletin* I, p. 11.

18 "IAEA Safeguards Overview: Comprehensive Safeguards Agreements and Additional protocols.".Available at http://www.iaea.org/publications/Factsheets/English/sg_overview.htm.

19 *Ibid.*

20 Russell Leslie et al, "Ensuring Effective safeguards Coverage of States with small Quantities protocol," p. 3. Available at http://www.dfat.gov.au/asno/publications/ensuring_effective_safegaurds_coverage_state_small_quantities_protocol.pdf (accessed on 11 July 2013)

21 Pierre Goldschmidt, 2001, "Strengthened Safeguards: Meeting Present and Future challenges," *IAEA Bulletin*, 43/4, pp. 9-10.

22 Quoted in Paul K. Kerr, December 15, 2011, "US Nuclear Cooperation with India: Issues for Congress," *CRS Report for*

Congress, Congressional Research Service, p. 10.

23 Quoted in "Government's Move a Shocking Betrayal: Left." Available at http://www.reiff.com/news/2008/jul/10ndeal8, (accessed on 22 September 2012).

24 R. Ramachandran, 16-29 August 2008, "Inching Closer," *Frontline*, vol. 25, issue 17.

25 Quoted in Siddharth Varadarajan, 2008, "Most IAEA Members Worried that Precedent being Set for Pakistan," *The Hindu*, 1 August.

26 Siddharth Varadarajan, 2008, "IAEA Safeguards Grant India Nuclear Weapon Status," *The Hindu*, 24 July.

27 Ed Markey quoted in "India - IAEA Safeguards Agreement Worse than Useless, A Shame," July 11, 2008. Available at http:/rediff.com/news/2008/july/11ndeal2.htm, (accessed on 17 September 2012).

28 Praful Bidwai, July 11, 2008, "India/US: Deceptions Surface in Nuclear Deal." Available at http://www.ipsnews.net/2008/07/india-us-deceptions-surface-in-nuclear-deal/ , accessed on 23 October 2012.

29 Gretchen Smith and Rekha Chakravarthi, July 26, 2008, "India-Specific Safeguards Agreement: India and American Responses," *Mainstream*, vol. XLVI, no. 2.

30 Jyantha Dhanapala and Daryl Kimbal, July 10, 2008, "A Non-Proliferation Disaster." Available at www.carniewgeendowment.org/2008/07/10/nonproliferation-disdaster/5e8271912012, (accessed on 2 October 2012).

31 *Ibid.*

32 Quoted in Praful Bidwai, "India/US Deceptions Surface in Nuclear Deal," *op. cit.*

33 "On IAEA Safeguards," August 2, 2008, Statement of Politburo of Communist Party of India (Marxist).

34 Vinay Kumar, 2008, "Nothing Unexpected in IAEA Safeguards Agreement: Karat," *The Hindu*, 12 July.

35 "IAEA Safeguards Agreement is Against India's Interests," July 26, 2008, The Left Parties Statement on IAEA Safeguards Agreement , *People's Democracy*, vol. XXXI, no. 28.

36 Vinay Kumar, 2008, "Nothing Expected in IAEA Safeguards, Karat," *The Hindu*, 12 July.

37 "BJP Questions on IAEA Draft," 2008, *The Hindu*,12 July

38 "BJP Lampoons Manmohan Singh on Proposed IAEA – India Specific Safeguards," July 14, 2008. Available at www.news.oneindia.in/2008/07114/bjp/-lampoons-manmohan-proposed-iaea-india-specific-safeguards-121090400htm, (accessed on 27 September 2012).

39 "Agreement between Government of India and the International Atomic Energy Agency for the Application of Safeguards to Civilian Nuclear Facilities," p. 2. Available at http://www.isis-online.org/publications/southasia/India_+IAEA_Safeguards.pdf, (accessed on 12 September 2012)

40 "IAEA safeguards Agreement is Against India's Interests," *op. cit.*

41 Brahma Chellaney, "The India-IAEA Safeguards Agreement Fact Sheet: Let the Facts Speak Themselves," July 12, 2008. Available at www.Chellaney.net/2008/07/12/india-iaea-safeguards-agreemet-factsheet, (accessed on 13 September 2012).

42 *Ibid.*

43 Prabir Purkayastha, 2008, "Does the IAEA Agreement Hide us From Hyde Act," *The Hindu*, 14 July.

44 Quoted in Siddharth Varadarajan, 2008, "Perpetuity of Safeguards Only with Perpetuity of Fuel Supply," *The Hindu*, 13 July.

45 Quoted in *Ibid.*

46 "Agreement between Government of India and the International Atomic Energy Agency for the Application of Safeguards to Civilian Nuclear Facilities," *op. cit.*

47 Quoted in Siddharth Varadarajan, "Perpetuity of Safeguards Only With Perpetuity of Fuel Supply," *op. cit.*

48 R. Ramachandran, " Inching Closer," *op. cit.*

49 Quoted in "On IAEA Safeguards", Statement of Politburo of CPI (M), *op. cit.*

50 Paul K. Kerr, "US Nuclear Cooperation With India: Issues for Congress," *op. cit.*, p. 12.

51 Siddharth Varadarajan, July 14, 2008, "Termination of Safeguards on Indigenous Reactors: What Gov/1621 Really Says." Available at www.Svardarajan.blogspot.in/2008/07/termination-of-safeguards-on-indigenous.htm., (accessed on 6 September 2012).

52 "Agreement between Government of India and the International Atomic Energy Agency for the Application of Safeguards to

Civilian Nuclear Facilities," *op. cit.*, p. 7.

53 *Ibid.*, pp. 8-9.

54 "GOV/1621 of 20 August 1973."Available at: www.armscotroll.org/system/files/gov1621.pdf, (accessed on 2 October 2012).

55 *Ibid.*

56 For details see Brahma Chellaney, "The India-IAEA Safeguards Agreement Fact Sheet: Let the Facts Speak Themselves," *op.cit.* and "India Safeguards Agreement Sinks." Available at www.livis.arscontrlwirk.com/archive/19447-india-safeguards-agreement-sinks.

57 Siddharth Varadarajan, July 16, 2008, "India's IAEA Draft: Invoking Para 29 won't be Easy." Available at www.rediff.com/news/2008/jul/16guest.htm, (accessed on 1 January 2013).

58 *Ibid.*

59 For details see R. Ramachandran, July 19 - August 1, 2008, "Open Secret," *Frontline*, vol. 25, issue 25 and Brahma Chellaney, "The India-IAEA Safeguards Agreement Fact Sheet: Let the Facts Speak Themselves," *op. cit.*

60 Cited in Prabir Purkayastha, "Does the IAEA Agreement Hide us from Hyde Act," *op. cit.*

61 *Ibid.*

62 Peter Crail, "AEA Approves India's Additional Protocol." Available at www.armscontrol.org/oct/2009_04/indiaprotocol, (accessed on 7 August 2012).

63 Siddharth Varadarajan, 2009, "India and The Additional Protocol," *The Hindu,*7 March.

64 *Ibid.*

65 *Ibid.*

66 Information Circular 254/Rev. 11/ Part I, November, 12, 2012.

67 Information Circular 254/ Rev. 8/ Part II, June 2010.

68 R. Ramachandran, September 27 - October 10, 2008, "Nuclear Hurdles," *Frontline*, vol. 25, issue 20.

69 "Nuclear Suppliers Group (NSG)." Available at: http://cns.miis.edu/inventory/pdf, (accessed on 12 June 2013)

70 Praful Bidwai, "Nuclear Waiver Blow to Non-Proliferation." Available at http://www.ipsnews.net/print.asp?idnews=43818, (accessed on 2 March 2013).

71 Quoted in Siddharth Varadarajan, 2008, "Waiver Enables Member States to Provide India Full Civil Nuclear Cooperation," *The Hindu*, 7 September.

72 Indian External Affairs Minister quoted in Jeffry Lewis, September 7, 2008, "India Gets NSG Waiver." Available at http://lewis.armscontrollwork.com/archive/2008/nuclear-deal-passes-nsg, (accessed on 20 March 2013).

73 Quoted in John Cherian, September 27 - October 10, 2008, "The Morning After," *Frontline*, vol. 25, issue 20.

74 See the "Text of the India-related waiver by the Nuclear Suppliers' Group (NSG)." Available at http://chellaney.net/2008/09/08/text-of-the-inda-related-waiver-by-the-nuclear-suppliers-group-nsg/, (accessed on 6 April 2013).

75 *Ibid.*

76 Quoted in "NSG Waiver Enables India to Sign Agreements with Countries: Pranab," 2008, *South Asian Times*, 6 October

77 Quoted in *Ibid.*

78 Quoted in "A unique Development says, Pranab," 2008, *The Hindu*, 7 September.

79 T.B. Subramanian, 2008, "A Breakthrough, says Kakodkar," *The Hindu*, 8 September.

80 "NSG Waiver: Another Surrender," September 7, 2008, Statement of Communist Party of India (Marxist)

81 "Condition of the Hyde Act Remain: Karat," 2008, *The Hindu*, 7 September.

82 Neena Vyas, 2008, "India has Walked into a Trap: BJP," *The Hindu*, 7 September.

83 "Tighten Draft Waiver for India, NSG Urged," 2008, *The Hindu*, 16 August.

84 Gretchen Smith, September 15, 2008, "India-NSG Waiver: Indian and American Responses." Available at http://www.ipcs.org/article/nuclear/indias-nsg-waiver-indian-and-american-responses-2681.htm, (accessed on 22 March 2013)

85 *Ibid.*

86 *Ibid.*

87 Praful Bidwai, "Nuclear Waiver: Blow to Non-Proliferation," *op. cit.*

88 R. Ramachandran, "Nuclear Hurdles," *op. cit*

89 Quoted in Malik Qasim Mustafa, "The Indo-US Civil Nuclear Deal:

An overview of IAEA Safeguards and Nuclear trade with NSG," A paper presented at a public Seminar on 24 September, 2008.

90 Daryl Kimbal, September 6, 2008, "Text, Analysis and Responses to the NSG Statement on Civil Nuclear Cooperation with India." Available at http://www.armscontroll.org/node/3345, (accessed on 7 July 2012)

91 See the "Text of India-Related Waiver by Nuclear Suppliers Group (NSG)," *op. cit.*

92 Siddharth Varadarajan, 2008, "New NSG Draft Has Small but significant Changes," *The Hindu*, 5 September.

93 John Cherian, " The Morning After," *op .cit.*

94 Siddharth Varadarajan, "Waiver Enables Member States to Provide India Full civil Nuclear Cooperation," *op. cit.*

95 Brahma Chellaney, August 15, 2008, "Serious implications for India in the NSG Draft." Available at http//www.redif.com/news/2008/aug/15brahma.html, (accessed on 4 April 20102).

96 Andrea Viski, May 2012, "The Revised Nuclear Suppliers Group Guidelines: A European Union Perspective," p. 6. Available at www.nonproliferation.en/documents/nonproliferation-nationpapers/andreaviski47ba1277ab879pdf, (accessed on 3 March 2013).

97 Quoted in *Ibid.*, p. 5.

98 R. Ramachandran, August 30 - September 12, 2011, "On Slippery Ground," *Frontline*, vol. 28, issue 16.

99 INFCIRC 254/ Rev. 9/p I

100 *Ibid.*

101 "NSG Imposes More Stringent Conditions for E and R Transfers," p. 1 Available at www.nei.org/fillefolder/IssueSumaaryNSG Guidelines.pdf, (accessed on 22 April 2013).

102 Quoted in "India Wants NSG Partners to Fully Honour Their commitments," 2011, *The Hindu*, 11 August.

103 Quoted in "New NSG Guidelines Limit India's Access to Sensitive Nuclear Technology." Available at http:www.wiseinterational.org/node/3759, (accessed on 22 November 2012)

104 Cited in Aditya Kaul, June 27, 2011, "Nuclear suppliers Group Revises Guidelines, India Wary." Available at http;//www.dnaindia.com/india/1559498/report-revises-guuidelines-india-wary, (accessed on 2 February 2013).

105 *Ibid.*

106 Quoted in Michel Krepon, 2011, "Another Contentious Nuclear Issue," *The Hindu, 12 July.*

107 Cited in Vladimir Radyuhin, 2011, "New ENR Norms Won't Hit Waiver to India: Russia," *The Hindu, 16 July*

108 Daniel Horner, July/August 2010, "NSG Makes Little Headway at Meeting." Available at http://www.arms.org/net/2010_07/NSGMtg, (accessed on 5 May 2013)

109 Vladimir Radyuhin, 2011, "Russia Silent on Whether NSG Ban will Cover India," *The Hindu*, 30 June.

110 Sandeep Dikshit, 2011, "Russia Alone Proposed a Way out of Tightened NSG rules, says Envoy," *The Hindu*, 8 December.

111 Cited in Sachin Parashar, 2011, "Now France Reaffirms NSG Rules Won't Affect Ties," *Times of India*, 2 July.

112 R. Ramachandran, "On Slippery Ground," *op. cit.*

4

Indo-US Civil Nuclear Deal: Issues And Implications

INTRODUCTION

The Indo-US Civil Nuclear Cooperation Agreement has become a subject of contentious debate among non-proliferation advocates in general and strategic experts in particular. The advocates of non-proliferation regime perceive the Agreement as a great challenge for global non-proliferation regime because the Agreement does not put a ceiling on India's fissile material production and provides India all the rights of civil nuclear technology that otherwise have been reserved for NPT signatory countries. However, in India, a different argument has surfaced. India's nuclear scientific establishment perceive the Agreement as a great challenge for India's strategic nuclear programme, placing unnecessary limitations on India's ability to secure a credible nuclear deterrent. There is another serious issue linked with the Agreement i.e.

regional implications of the Agreement which largely emanate from the prevailing security dilemma between India and Pakistan on one hand and between India and China on the other hand. Pakistan has already expressed its deep concerns over the Agreement that has presumably tilted the strategic balance of the region in favour of India and has warned of taking every possible measure to maintain a strategic balance in the region.

This chapter has been divided into four main sections. Section one, analyses the implications of the Agreement for global non-proliferation regime. Section two, deals with the implications of the Agreement for India's nuclear weapons programme. Sections three, discusses the regional ramifications of the deal and section four, deals with the issue of India's energy security and the possible role nuclear energy could play in meeting India's long term energy needs.

INDIA-US NUCLEAR PACT AND NUCLEAR NON-PROLIFERATION REGIME

One of the central and contentious issues linked with India-US Civil Nuclear Cooperation Initiative is what will be the implications of the Agreement for nuclear non-proliferation regime: does it promote or undermine the non-proliferation regime? In fact, at the very first hearing on nuclear deal in US Congress, Henery J. Hyde noted that "among the critical questions surrounding this agreement was whether its net impact on our non-proliferation policy is positive or negative."[1] Significant number of Congressmen and experts who testified before different Congressional Committees argued that the pact undermines US and global non-proliferation efforts and viewed it a great challenge for global non-proliferation regime. It was argued that India is an NPT outlier and weakening of global rules for it would create a breach in global export control system letting the rogue states to drive through and this could have far reaching consequences for US and global non-proliferation policy. In the words of Gary Milhollin: "It is impossible to weaken export controls for India without weakening them for everyone else. The "everyone" includes Iran, Pakistan and even terrorists.... Indeed, if Washington does weaken export controls for everyone, which is bound to happen if it weakens them for India, it may hasten the day when a nuclear explosion destroys a US city."[2] Proponents of the Agreement however argue that the Agreement is a major incentive to bring India into global non-proliferation regime and civil nuclear cooperation initiative is a major reward for India's "responsible" behaviour and impeccable non-proliferation record.

PROPONENTS VERSUS OPPONENTS

What Opponents Say

The nuclear pact unfolded on 18 July 2005 by US President George W. Bush and India Prime Minister Dr. Manmohan Singh is contrary to US non-proliferation policy and quite damaging to global non-proliferation efforts. This is the view largely shared by non-proliferation ayatollahs world over. The deal has been largely criticised on the grounds that it provides the benefits of civil nuclear technology to India which have been reserved for NPT signatory countries. It fails to cap India's fissile material production and undermines the efforts to prevent rogue states from acquiring nuclear weapons. Launching a direct attack on the nuclear deal, Congressman Edward J. Markey on the occasion of roll call vote said that:

> *"With this vote, we are shattering the non-proliferation rules and the next three countries to march through the broken glass will be Iran, North Korea and Pakistan...[This] is an*

absolutely crazy situation for us to be engaging in...this is ripping [NPT] foundation up by its roots."[3]

The efforts of the Bush administration to pursue the deal received a great set back when eighteen experts on non-proliferation and security matters wrote a letter to US Congress on November 18, 2005, urging the Congress to impose additional obligations on India before considering the proposal. Signatories noted that the proposal poses far reaching consequences for US non-proliferation policy and does little to bring India into non-proliferation regime. Commenting on the non-proliferation risks associated with the deal, experts underlined that India has agreed to accept voluntary safeguards on its nuclear facilities and this would allow India to remove its facilities from safeguards on national security grounds. The signatories further said that India should be asked to bring its current and future nuclear facilities under permanent safeguards to ensure that there is no diversion of imported nuclear fuel. Signatories also urged that no nuclear assistance should be provided to India unless it stops the production of fissile material.[4]

A serious lacuna of India-US Civil Nuclear Cooperation Initiative according to non-proliferation advocates is that it does not require India to stop its production of fissile material and it fails to bring India's huge stockpiles of spent nuclear fuel under safeguards. It is estimated that over the years, India's unsafeguarded nuclear reactors may have generated spent fuel containing around 9,000 kilograms of un-separated plutonium. With this plutonium India can produce around 1,100 nuclear weapons.[5] Another serious loophole of the Agreement is its failure to capture a limited number of India's nuclear reactors for safeguards. As per 2006 "separation plan", India has agreed to place fourteen out of twenty-two reactors under safeguards. The eight unsafeguarded reactors currently used for power production could be used for the production of weapons grade plutonium. When operated at 60-80 per cent capacity factor and by limiting the time the fuel is irradiated, a 220 MW reactor could yield 150-200 kilograms of plutonium annually.[6] If operated in this fashion, India's eight unsafeguarded reactors could produce fissile material sufficient for hundreds of nuclear warheads annually. Moreover India has decided to keep its breeder reactors outside the purview of international monitoring to serve its strategic nuclear programme.

The above argument leads us to a very important question: what will be the impact of the Agreement on India's nuclear weapons programme and how will it influence the behaviour of NPT signatory countries. It has been argued that the biggest advantage that nuclear deal provides to India is that it would free India's domestic reserves of uranium for the production of nuclear weapons as India could use imported fuel to feed safeguarded reactors and divert indigenous reserves towards weapons programme. To quote K.Subrahmanyam, former head of National Security Advisory Board:

> *"Given India's uranium ore crunch and the need to build up a minimum credible nuclear deterrent arsenal as fast as possible, it is to India's advantage to categorize as many power reactors as possible as civilian ones to be refuelled by imported uranium and conserve our native uranium for weapon grade plutonium production."*[7]

A former Indian intelligence official has claimed that by freeing India's domestic reserves of uranium nuclear deal would enable India to produce at least fifty nuclear weapons annually.[8] Thus instead of containing India's nuclear weapons programme, nuclear deal promotes it, which is major non-proliferation loss.

Another downside of the Agreement as per non-proliferation priests is its adverse impact on NPT signatory countries like Brazil, South Africa, Japan etc, who in spite of having technological capability renounced the nuclear weapons path under this promise that the benefits of civil nuclear technology will be reserved for NPT signatory countries only. However, because of the Bush-Singh initiative which provides India the rights of civil nuclear trade otherwise reserved for NPT signatory countries only, these countries will "regard NPT as an anarchism, reconsider their self restraint and be tempted by the precedent that India has successfully established and that in fact has an American blessing."[9] This point was stressed by Egypt's (UN) Ambassador Maged Abdul Aziz at 2010 NPT Review Conference: "If you say countries outside the treaty are going to get ...even more benefits than countries inside the treaty, then what is benefit for me to bind myself with more [non-proliferation] restrictions."[10] The precedent set by the deal will manifest itself in similar deals floated by advanced countries for their strategic partners. A glaring example is the nuclear deal between China and Pakistan signed in 2010 for the construction of two nuclear reactors in Pakistan. Israel has also shown a great interest to ink a similar type of agreement with US and has made a formal request for the same. The deal also provides an incentive to other countries considering to pursue nuclear weapons path. The Agreement provides them enough reasons to believe that nuclear weapons are power maximisers and if they acquire nuclear weapons they will be slapped with sanctions for a period of time but ultimately sanctions will be lifted and they will be admitted to the nuclear club like India.[11]

The NSG waiver granted to India in September 2008

is also likely to have negative repercussions. NSG is a unique cartel which operates on "consensus principle". Any decision taken by the cartel has to be agreed by all the members of the group. In its September 8 decision, all the members agreed to relax the guidelines to accommodate India but they failed to lay down the penalties in case India violates its non-proliferation commitments. If such an eventuality takes place in future, all the members of the group will meet and take a decision based on consensus to decide future course of action against India but in such a scenario it is highly likely that profit taking will triumph non-proliferation gains. Russia and France, two great beneficiaries of the initiative will evade the consensus and would like to continue their nuclear business with India at the cost of non-proliferation principles.[12]

Non-proliferationists also criticised Bush administration's approach to reverse the fundamental credentials/principles of US non-proliferation policy merely for commercial gains. It has been decades old US policy to limit the number of nuclear weapon states only to five but by signing the civil nuclear agreement with India which is not member of NPT and has an active nuclear weapons programme, the administration has accorded a de facto nuclear weapon state status to India. Under NPT, US is obligated not to provide any sort of assistance- direct or indirect to any country which can fuel its nuclear weapons programme. Moreover, US domestic laws forbid cooperation in the field of nuclear energy with any state that does not have a Comprehensive Safeguards Agreement in place. But Bush administration has reversed this decades old US policy merely to score commercial gains which remain elusive. Even after eight years of the signing of joint statement on civil nuclear cooperation, US companies are still struggling to complete a single reactor sale with India. The prospects of future US engagement in Indian nuclear market also remain bleak due to tough competition posed by Russian and French nuclear companies and benefits and opportunities created by US are likely to go elsewhere France and Russia.[13]

Besides non-proliferation advocates, a number of countries also criticised the Bush proposal for its adverse impact on non-proliferation regime. Deputy Head of Iran's Atomic Energy Organisation Mohammad Saeedi said that the deal will trigger "new crises" in international community. The transfer of nuclear technology to NPT outliers will endanger NPT.[14] Ali Larijani, Iran's chief nuclear negotiator charged the US of pursuing "dual standard" by continuing its relations with India which has not signed NPT and warned that such a "dual standard" would prove detrimental to global security.[15] China's official news agency Xinhua reported that the nuclear pact with India "will set a bad example for other countries".[16] Peoples' Daily commented that the deal "will inflict a hard blow to global non-proliferation regime" as most of the states will follow the example set by US and cut similar type of deals with their allies.[17] Japan's Foreign Minister, Taro Aso, noted that because of the Agreement, NPT "would lose its substance" as India is not a signatory to the treaty (NPT) like US.[18]

To allay the concerns of non-proliferation lobby several amendments were moved in the Senate and House of Representatives but were voted down. In the House an amendment requiring US to withhold fuel supplies to India till it ceases the production of fissile material was moved by Representatives Howard Berman. Another amendment was moved by Representative Brad Sherman calling up on the president to certify annually that India has not increased the level of its domestic uranium used in its weapons programme during the preceding year in order to receive the fuel for current year. In Senate, at least three amendments of same nature were moved but were defeated.[19]

What Proponents Say

Contrary to the arguments of non-proliferationists, proponents of the deal argue that the deal is a major incentive to bring India within the ambit of non-proliferation regime and it is reward for India's "responsible" behaviour and exceptional non-proliferation record. In her testimony before the Senate, US Secretary of State Condoleezza Rice pleaded that: "Some have expressed concern that civil nuclear cooperation with India will weaken the NPT, or undermine global non-proliferation efforts. Dr. Mohamed ElBaradei, the Director General of IAEA ... does not share this concern, just the opposite. Dr. ElBaradei publically praised the initiative the day it was announced, stating that it will 'bring India closer as an important partner in the non-proliferation regime... it would be a mile stone, for ongoing efforts to consolidate the non-proliferation regime...'"[20] Proponents of the Agreement consider it a major non-proliferation gain. Since 1974, India was treated as a pariah in non-proliferation order and was subjected to nuclear isolation which provoked India to continue to its nuclear weapons programme. So, it was a great challenge for non-proliferation regime to bring India within the ambit of non-proliferation mainstream. To quote US under Secretary of State, Nicholas Burns:

> *"We sought this agreement because India's nuclear weapons programme and its status outside the non-proliferation has proven to be a longstanding stumbling block to enhanced US-India relations as well as a problem for global non-proliferation regime. The initiative for civil nuclear cooperation*

announced by President Bush and Prime Minister Singh in July is intended to deepen the bilateral partnership...and advance international non-proliferation norms and practices... Although India has demonstrated a strong commitment to protecting fissile materials and nuclear technology more generally, it is in both Indian and American Interests that New Delhi's isolation be brought to an end and that India be made a part of stable non-proliferation order. The agreement between President Bush and Prime Minister Manmohan Singh does this in a fair and equitable way. It contemplates both the countries taking serious steps towards achieving the goal of strengthening the international non-proliferation regime..."[21]

The benefits of bringing India in to global non-proliferation regime were even forcefully stressed by Ashley J.Tellis of Carnegie Endowment for International Peace, who said that India's entry in to non-proliferation regime would not only strengthen global non-proliferation standards but will also enhance American national security. Tellis even challenged the argument that the deal will open the door for other states asking for similar treatment. He said that India is a unique case and it must not be treated at par with other states that have often violated their international commitments. India is a responsible state and deserves a special treatment.[22]

Proponents of the Agreement claim that the non-proliferation gains will largely flow from the commitments made by Indian Prime Minister in July 18 joint communiqué. In compliance to these commitments, India has declared fourteen of its reactors as civilian which will be placed under the safeguards of IAEA. This will open sixty-five per cent of its reactors for international inspection. In addition to this, it has signed an Additional Protocol with IAEA and has also taken a pledge to continue its voluntary moratorium on nuclear testing, to strengthen nuclear export control system and to refrain from transferring sensitive nuclear technologies to other states that do not possess such technologies. All these are positive steps which will not only strengthen global non-proliferation practices and standards but will also promote global security.

What Facts Say

While going through the arguments put forth in favour and against the Agreement so far its implications for non-proliferation regime are concerned, it becomes obvious that some of the arguments put forth by "opponents" and most of the arguments held by "proponents' are unconvincing and overstated. In the case of "opponents", the argument that the initiative will aggravate the problem posed by nuclear aspirants like Iran and North Korea holds little weight. There is a broad consensus among strategic experts that the states pursue nuclear weapons on the basis of their own security calculations and in the case of Iran and North Korea, their nuclear programmes predate the India-US Civil Nuclear Initiative and it is highly unlikely that the initiative will drive them on weapons path. Prof. Stephen Cohen is of the opinion that there is hardly any evidence that the nuclear deal might have influenced Iran's decision on uranium enrichment: "I doubt it, but without better access to Iranian-decision making system, no one can really draw a conclusive judgement, certainly that the Iranians have for some time been using India-derived rhetoric to trash the NPT..." [23] Similarly, the claim that India's Safeguards Agreement signed with IAEA is voluntary in nature and India can remove its facilities from safeguards on national security grounds is highly exaggerated. India has signed a facility-specific Safeguards Agreement with IAEA and termination of safeguards on any facility will be jointly determinate by IAEA and India.

In the case of "proponents", most of the arguments put forth are weak and can be contested on number of grounds. In the first place, the claim that the deal brings India into non-proliferation regime is highly overstated. The claim is largely based on the commitments made by Indian Prime Minster in July 18 joint communiqué that: India will continue its voluntary moratorium on nuclear testing, sign a Safeguards Agreement and an Additional Protocol with IAEA, strengthen its nuclear export control system and support the negotiations on a Fissile Material Cut off Treaty. It must be noted that most of these commitments are not the product of Civil Nuclear Cooperation Initiative and were in place before the negotiations on the initiative took off in 2005. India is observing a voluntary moratorium on nuclear testing since 1998 and is a part of India's nuclear policy since then. Strengthening of export control system is not the product of the deal either. In 2004, UNSC passed a resolution 1540 which called upon all the states to improve and strengthen their national export control systems to prevent the proliferation of nuclear weapons. It was in conformity with this resolution that India enacted a law in 2005, before the announcement of civil nuclear initiative to strengthen its nuclear export control system.[24] Still, India has a long way to go to achieve international standards.

India has been observing safeguards on a number of its nuclear facilities from a very long time and signing of a new Safeguards Agreement is a welcome move but its importance from non-proliferation stand point is only symbolic. India-IAEA Safeguards Agreement captures only fourteen of India's twenty-two reactors. The remaining eight reactors will serve India's strategic nuclear programme. Moreover, India's breeder programme is also out of safeguards. The breeder reactors along with eight unsafeguarded reactors can produce fissile material sufficient for hundreds of warheads annually. So, if India runs a nuclear weapons programme parallel to its civilian nuclear programme the notion of safeguards is misleading.

India's pledge to negotiate a Fissile Material Cut off Treaty is not going to break any new ground either. India knows that negotiating such a treaty will take years before it reaches to completion. Moreover, India has differences with US on some of the provisions of the treaty. India is in favour of a verifiable cut off treaty but US does not support a verifiable mechanism. This difference between the two can further derail the process of negotiations.

Secondly, proponents of the Agreement claim that India has an impeccable non-proliferation record and the deal is a reward for its exceptional record. But India's non-proliferation record is not a clean as claimed by the proponents. It is a known fact that in its 1974 nuclear test India diverted plutonium form 40 MW reactor supplied by Canada. The heavy water for the reactor was supplied by US. It was a point raised by Senator Richard Lugar in his opening statement: "India's nuclear record with international community had been unsatisfying and that India had violated bilateral pledges it made to Washington not to use US supplied nuclear material for weapon purposes."[25] European "early warning" intelligence assessment of July 2005 lists India as one of the six proliferator countries because of illicit procurement of nuclear equipments for its nuclear programmes. The report warns the European companies about Indian entities that are targeting European companies to buy material for weapons programme.[26]

The July 18 joint communiqué describes India as a "responsible state" with advanced nuclear technology and India has also reiterated to "assume same responsibilities and practices and acquire same benefits and advantages as other leading countries with advanced nuclear technology," but still India has a long distance to cover to become a responsible nuclear weapon state. All the responsible states have signed the treaty banning all nuclear weapon tests and have ceased the production of fissile materials but India has not undertaken these obligations yet. Responsible behaviour comes by undertaking responsibilities, not by avoiding them.

NUCLEAR DEAL AND INDIA'S NUCLEAR WEAPONS PROGRAMME

The India-US Civil Nuclear Cooperation Agreement which came under heavy attack in US from non-proliferation lobby on the grounds that the Agreement does little to cap India's growing nuclear arsenal and assists India's nuclear weapons programme also came under a sharp criticism from certain sections of Indian nuclear scientific and strategic establishments on the grounds that the Agreement is detrimental to India's strategic nuclear programme. They see the Agreement imposing unnecessary restrictions on country's nuclear programme and comprising its quest for a 'credible nuclear deterrent'. According to them, different arrangements and commitments made by the Indian government as a part of the Agreement pose a great challenge to India's strategic weapons programme including potential weakening of India's nuclear deterrent and erosion of indigenous research and development. These arrangements and commitments include:

- India's "separation plan".
- India-IAEA Safeguards Agreement.

- Commitment to negotiate a Fissile Material Cut off Treaty.
- Voluntary moratorium on nuclear testing.

1. Separation Plan

In March 2006, during the visit of American President George W. Bush to India, two sides agreed to an arrangement to divide India's nuclear programme into military and civilian ones. Under the arrangement India agreed to place fourteen reactors under civilian list along with a number of upstream and downstream nuclear facilities. As a part of the arrangement India also agreed to shut down CIRUS reactor in 2010 and move core of the Apsara reactor out of BARC and to place it under safeguards. A number of former and serving member's of Indian scientific establishment took a strong exception to the Indian decision. Reacting to the decision, former Director of BARC, A.N. Prasad said:

> *"Segregation of civilian and military facilities in the nuclear field is 'impossible'. Our military activities are not aimed at stockpiling nuclear weapons. Rather the aim is deterrence, which in turn is based on a given level of 'threat perception'. Since the Unites States and other big nuclear weapon states have doctrines based on stockpiling, they can perhaps afford to maintain dedicated military facilities for the production and maintenance of nuclear munitions. 'But even they are finding that stockpiling imposes further costs...' For India ... going down the route of stockpiling... would be 'highly counterproductive' and costly. Separating the civilian from nuclear, as the Prime Minister has committed the country doing means having 'declared, dedicated facilities for the military side...'*

> *Having dedicated facilities will terribly raise the cost of the weapons programme'."*[27]

Former Indian National Security Adviser, Brijesh Mishra termed the "separation plan" "an unfortunate, unwelcome outcome for India" which could have adverse impact on India's nuclear weapons programme.[28] The decision to place eight additional Pressurised Heavy Water Reactors under safeguards and to shut down CIRUS reactor in 2010 also drew a sharp criticism. It has been argued that to close down the CIRUS reactor, a key source of India's weapons grade plutonium without a replacement will largely hurt India's strategic nuclear weapons capability. Another crucial issue linked with the "separation plan" is tritium – an important material required for the development of thermo-nuclear weapons. It is produced as a by-product in PHWRs. Tritium has a very short lifespan and decays quickly as compared to other nuclear weapon materials, so it has to be produced on regular basis to balance its decay. Indian strategic analysts fear that the decision to place fourteen reactors in the civilian list will limit India's tritium inventory to a lower level which in turn will affect its ability to produce high yield thermo-nuclear weapons for credible national deterrent.[29]

In response to the criticism levelled against the "separation plan," Prime Minister Manmohan Singh stated in the Parliament that:

> *"Separation plan will not adversely affect our [India's] strategic programme. There will be no capping of our strategic programme and the separation plan ensures adequacy of fissile material and other inputs to meet the current and future requirements of our strategic programme, based on our threat scenarios. No constraint has been placed on*

our right to construct new facilities for strategic purposes. The integrity of our Nuclear Doctrine is adequately protected. The separation plan does not come in the way of the integrity of our three-stage nuclear programme, including the future use of our thorium reserves. The autonomy of our research and development activities in the nuclear field will remain unaffected."[30]

A close perusal of the "separation plan" confirms the above statement of Prime Minister Dr. Manmohan Singh and gives the impression that it does not affect India's strategic nuclear programme and leaves its nuclear weapons capability intact. Firstly, the decision to shut down the CIRUS reactor was taken because of two reasons, one, that it has always been a source of great controversy between India and the US and there was strong pressure from US non-proliferation priests to place it under safeguards. Second, it was located at BARC, a site of great strategic importance for India. So, the decision to shut it down instead of placing it under safeguards will not only keep the BARC off the sight of international inspectors but will also keep silent the US non-proliferation priests.

Secondly, placing eight additional reactors in the civilian list will not affect the size of India's nuclear arsenal; rather it will have a reverse affect. Placing reactors in the civilian list means that the fuel for these reactors will come from outside and this will free India's indigenous reserves of uranium which it can devote to its weapons programme.

Thirdly, the research facilities which have been placed under the civilian list are not relevant for safeguards. Great care has been taken to ensure the integrity and autonomy of India's three-stage nuclear programme. A number of important enrichment and reprocessing facilities associated with nuclear fuel cycle have been kept in the military list.

Lastly, India does not require its power reactors to maintain its tritium inventory. There are alternate ways to produce it. India can use its detritiation plants as well as its research reactor Dhruva to maintain its stockpile of tritium required for credible nuclear deterrent. [31]

2. IAEA Safeguards

The issue of safeguards is a subject on which a great deal of controversy surfaced in the country. In the July 18 Joint Communiqué Prime Minister Dr. Manmohan Singh has recorded that India will negotiate a Safeguards Agreement with IAEA and will place the civilian part of her nuclear programme under IAEA safeguards. Indian scientific establishment was greatly alarmed by the decision and expressed deep reservations vis-à-vis the decision. It was argued that international safeguards will greatly hamper the technological innovation and development of India's nuclear programme and will seriously erode the autonomy of India's indigenous research and development programme. Once India will open its nuclear programme to IAEA inspectors "[a]n international 'licence e-permit raj' on Indian scientific creativity will be here to stay and the army of IAEA inspectors will invade all related public and private sector entities, sometimes even without prior intimation. At the very least, it would guarantee that scientists and engineers would be endlessly tied up in bureaucratic red-tape so as to satisfy an infinite number of queries so that very little constructive work is achieved."[32] Commenting on the adverse affects of international safeguards Dr. A.N. Prasad said that safeguards are detrimental to research and development, "Safeguards are like cancer. Once they get into the system, they spread through under the 'pursuit and contamination provisions' of safeguards agreements, which can't be avoided and also are not amenable for negotiations".[33] According to P.K. Iyengar, former Chairman of India's Atomic Energy Commission, safeguards will erode the autonomy of India's indigenous research and development as IAEA inspectors will frequently visit these research facilities and Indian scientist will always find them looking over their shoulders. After experiencing same inconvenience Brazil, an NPT signatory country has objected to IAEA inspections on her domestic projects.[34]

However, keeping in view the "separation plan" agreed between two countries and nature of the India-IAEA Safeguards Agreement it seems unlikely that there will be any sort of interference in India's strategic nuclear programme on the part of IAEA inspectors in the name of safeguards. By creating a wall of separation between its civil and military nuclear programmes, India has left one part of its nuclear programme open for international inspections and IAEA has no mandate to see what is going on other part. The Agreement reached between India and the Agency does not give it a license to interfere with India's strategic nuclear programme which has been kept off the safeguards. India can out-rightly refuse any inspection or intrusion which it feels is beyond the mutually agreed arrangement.

Another issue linked with safeguards is India's fissile material production. It has been mentioned that under the Safeguards Agreement India will open eight additional reactors for international inspections which means that these reactors will become unavailable for military use. Some scientists allege that it will affect India's fissile material production and will constrain India's nuclear weapon capability. This argument however misses the point that India has enough stock of fissile material and it does not need these reactors to meet the requirements of a credible deterrent and enough options have been left open to meet future requirements. As per the estimates of a research report of International Panel on Fissile Materials, by September 2006 India had accumulated about 500 kg of weapons grade plutonium from its two research reactors, CIRUS and Dhruva. We can add another 45 kg of weapons grade plutonium to this figure which CIRUS would have generated before its decommissioning in 2010. Dhruva annually produces about 20-25 kg of weapon grade plutonium.[35] Taken together these figures and the rate of production of fissile material by Dhruva it can be estimated that India's stock of weapons grade plutonium by the end of 2013 would be around 685-720 kg. With this stock, Indian can roughly produce 137-144 war heads. India has also kept its 500 MW PFBR out of safeguards. This reactor is under construction. It could annually produce 130 kg of weapons grade plutonium.[36] Besides the weapons grade plutonium, India has around 11.5 tons of reactor grade plutonium in the spent fuel generated by its unsafeguarded reactors of which around 9 tons have been separated, the report claims. Under the terms of the deal India has agreed to place eight additional reactors under safeguards, these reactors would add another 4 tons to the stock of 11.5 tons, before they are placed under safeguards. [37] With this stock, India can produce over 1000 nuclear weapons. Some members of India's nuclear scientific community have expressed a concern that this reactor grade plutonium is unsuitable for the development of nuclear weapons as it has a different mix of isotopes than weapons grade plutonium. However, it is worth mentioning that in 1962, US conducted a nuclear test involving reactor grade plutonium. Also, in one of India's 1998 nuclear rests reactor grade plutonium was used.[38] No doubt the "separation plan" will lead to 1/3rd reduction in India's annual plutonium production but once 500 MW

PFBR will become operational there will be four to five fold increase in India's weapons grade plutonium production capacity.[39]

Table 4.1. Projected plutonium production from 2007 till reactors are safeguarded

Reactor	Proposed Date of safeguarding	Plutonium production (kg) Before Reactor is safeguarded
Rajasthan -3	2010	475
Rajasthan-4	2010	475
Kakrapar-1	2012	712
Kakrapar-2	2012	712
Narora-1	2014	950
Narora-2	2014	950
Total		4274

(*Source:* Fissile Materials in South Asia: The Implications of U.S.-India Nuclear Deal, A Research Report of International Panel on Fissile Materials)

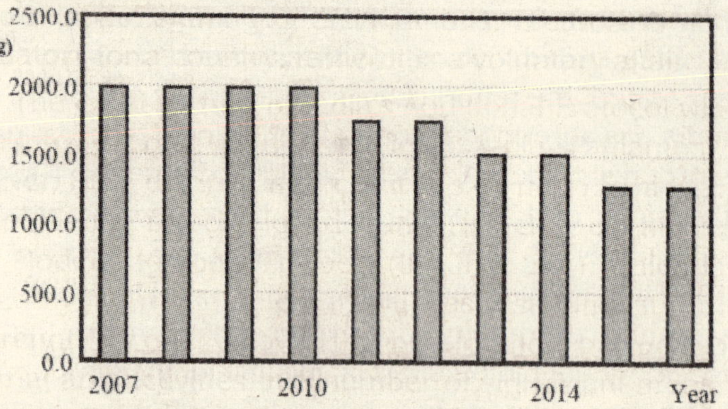

Fig. 4.1. Annual production of unsafeguarded plutonium from all Indian power reactors from 2007 until 2016, as reactors are progressively placed under safeguards.

(*Source:* Fissile Materials in South Asia: The Implications of U.S.-India Nuclear Deal, A Research Report of International Panel on Fissile Materials)

3. Fissile Material Cut off Treaty

July 18, 2005 joint statement notes a commitment

made by Indian Prime Minister Dr. Manmohan Singh that India will work with US towards the conclusion of a multilateral Fissile Material Cut off Treaty. Many Indian nuclear scientists and strategic analysts claim that this commitment will cap country's fissile material production and will lead to potential weakening of India's nuclear deterrent by limiting the size of country's nuclear arsenal. P. K. Iyengar, former Chairman of India's AEC considers it a "dangerous commitment", keeping in view India's need for fissile material, which is nowhere near the stock of five recognised nuclear powers and argues that it is "inconsistent" with India's policy of 'credible minimum deterrent' required for strategic purposes. [40] Strategic analyst Bharat Karnad argues that India's accession to Fissile Material Cut off Treaty will mark an end of India's fissile material production which could have far reaching consequences for India's nuclear weapons programme. It will make "irrelevant the "separation plan" and the eight power stations, the designed breeder reactors and the dedicated military reactor (100 MW Dhruva) and the rest of nuclear caboodle consigned to the "military" part of the nuclear energy programme."[41] Moreover, the commitment will push India into non-proliferation net and will serve the longstanding objective of US non-proliferation policy: "capping and freezing" of Indian nuclear programme.[42] However, Indian official position is that it will join a "non-discriminatory, multilaterally negotiated and internationally verifiable" Fissile Material Cut off Treaty provided that its security interests are fully addressed.[43]

A great strategic advantage for India is that the 123 Agreement per se does not require India to halt its fissile material production. The question of putting a ceiling on India's fissile material production will arise only when FMCT is negotiated in Conference on Disarmament (CD) that is highly unlikely in near future keeping in view the differences between the members of Conference on Disarmament. For instance Pakistan, which for many years has blocked the negotiations on FMCT is in favour of a treaty taking care of present, past and future production of fissile material as it considers itself in a disadvantageous position vis-à-vis India in terms of fissile material stock. India on the other hand favours a verifiable treaty banning only future production of fissile materials. Five nuclear weapon states like India want a treaty banning future production of fissile materials but have strong reservations vis-a- vis the verifiable mechanism. These differences cannot be easily bridged giving ample time to India to accumulate enough fissile material to satisfy the requirements of minimum credible deterrent.

4. Moratorium on Nuclear Testing

A crucial and a complex issue associated with India-US Civil Nuclear Cooperation Agreement is "India's right to test" a nuclear device. Some members of India's scientific and strategic communities allege that by agreeing to voluntary moratorium on nuclear testing in July 18 joint communiqué India has surrendered her right to test a nuclear device for all practical purposes. It is claimed that resuming 'nuclear testing' is necessary for India because of the failure of thermonuclear device tested in 1998 and limitations of country's simulation capability.[44] In order to achieve a credible, safe and reliable nuclear deterrent, India will have to resume testing; otherwise its nuclear deterrent will remain frozen at a primitive and low quality level. To quote strategic analyst Bharat Karnad:

> *"With the July 18 Joint Statement conferring the moratorium, India for all intents and purposes is permanently saddled with a half-baked nuclear deterrent "deployed" in a half cocked "de-mated" mode, which will deter*

and dissuade nobody and should its nuclear bluff be called in a crises, would become a liability."[45]

The history of India's voluntary moratorium on nuclear testing dates back to 1998 when Vajpayee government after Pokhran-II test series announced a voluntary moratorium on nuclear testing. In a background paper introduced in Indian Parliament, government announced that "[i]n terms of technical capability our scientists and engineers have the requisite resources to ensure a credible deterrent."[46] Thereafter in 2003 when India's nuclear doctrine was announced, government reiterated its pledge to continue its moratorium on testing and since then it is a part of India's nuclear policy and same has been reiterated in July 18 joint statement. However, what is important is that 123 Agreement which is binding on both the countries does not make any explicit reference to nuclear testing and the Agreement per se does not prohibit a nuclear test. But there is a concern emanating from US domestic law, Hyde Act. Section 106 of the Act states that "any waiver shall cease to be effective if the president determines that India has detonated a nuclear explosive device after the enactment of this title." So, if India reviews its voluntary moratorium policy and detonates a nuclear device then under section 106 of the Act US can cease its nuclear cooperation with India and under Article 14 of the 123 Agreement has a right to serve a notice of termination of the Agreement to India. If the Agreement is terminated, US under Article 14(4) of the Agreement can also seek return of nuclear material, equipment and components transferred under the provisions of the Agreement. However before the Agreement is terminated, it calls up on both the parties to hold consultations and take in to account the "circumstances that may lead to termination or cessation resulted from a party's serious concern about a changed security environment or as a response to a similar action by other states which could impact national security." [47] So, Agreement gives enough manoeuvring room to India to defend its nuclear test if her security interests are jeopardised because of a nuclear test conducted either by Pakistan or China. In such a scenario, US will have to withdraw notice of termination and to continue its nuclear cooperation with India. Besides the "security considerations", "preferences of US leaders, domestic lobbying and geopolitical balance of power"[48] are three factors which will also play a decisive role in determining US response to Indian nuclear test.

NUCLEAR DEAL AND ITS REGIONAL IMPLICATIONS

Any debate on the implications of the India- US Civil Nuclear Cooperation Agreement will be incomplete unless the broader strategic implications of the Agreement for South Asian region where India is experiencing strained relations with its traditional adversary Pakistan and is also facing a tough completion form China, which for years has emerged as a dominant player of the region are evaluated. Since Indo-US Civil Nuclear Deal is not merely a bilateral Agreement related to peaceful aspects of nuclear energy but it has a strategic intent and has the potential to influence to strategic landscape of the region; in this context it also becomes important to know how two important stakeholders of the region, Pakistan and China who share a strong strategic relationship which has been dubbed as " higher than mountains and deeper than oceans" are going to respond to emerging strategic entente between India and US. Pakistan has already expressed its deep concerns over the Agreement and has called it a "discriminatory" Agreement[49] having the potential to offset the strategic balance of the region.

Given the fact that Pakistan's defence and foreign policies have always been as Dr. Rasul Bukush says "India centred from the beginning" and "for over half a century Pakistan's security dilemma has centred on how to balance, counter and if necessary fight the Indian threat"[50] one can expect that Pakistan will possibly take every possible measure to maintain the strategic balance in the region as former Pakistani President Pervez Musharaf has asserted that "We remain oblivious to the changes evolving in the region. All the steps will be taken for the defense, security and safety of Pakistan".[51] China, that too is suspicious of growing defence and strategic cooperation between India and US is also likely to react to preserve its predominant position in the region by countering the growing influence of India.

Initially Pakistan's reaction to the India-US nuclear pact was very mild, possibly Pakistan was expecting a similar deal from Bush administration as offered to India. In a statement issued on 2 March 2006, the day when President Bush was in India to clinch a deal with India on "separation plan", Pakistan's foreign office spokesperson Taslim Aslam noted that "Pakistan has the same claim and expectation of international cooperation under safeguards for nuclear power generation, especially because Pakistan is a fossil fuel deficient country and has a significant and fully safeguarded nuclear power generation programme."[52] However, when President Bush ruled out the possibility of sealing a similar agreement with Pakistan on the grounds that Pakistan and India are two different countries "with different needs and different histories"[53] a significant change came in Pakistan's position. Not directly citing the security and risk of proliferation of nuclear materials as a reason for holding back nuclear cooperation with Pakistan, US cautiously played a different card as Richard Boucher, Assistant

Secretary of State for South and Central Asia underlined that:

> *"Our energy dialogue with Pakistan is going to be different from our energy discussions with India. One should not expect that (Pakistan's) energy needs would be met the same way, given different geography, different history and different resource base."*[54]

Realising that civil nuclear cooperation agreement with US along the lines as India has signed is a distant possibility Pakistan openly came against the India-US Agreement. In a statement Pakistan's foreign office noted that the Agreement would have serious implications for the South Asian security environment and global non-proliferation regime as Agreement could trigger an arms race in the region. Taking a strong exception to the failure of the Agreement to cap India's nuclear weapons programme the statement said that it would encourage India to continue its development of nuclear weapons which could prove detrimental to the stability of the region.[55] The deal even triggered a meeting of Pakistan's National Command Authority (NCA) to take stock of country's deterrence capability. The meeting which was attended by President Pervez Musharaf and other key officials of defense establishment expressed the concerns over India-US Nuclear Deal and reiterated the commitment to maintain Pakistan's credible deterrence capability.[56]

One of the major concerns of the Pakistan's strategic community is the failure of the Agreement to place restraints on India's nuclear weapons programme. The Agreement underlines that the purpose of "the Agreement is to provide for peaceful nuclear cooperation and not no affect the unsafeguarded nuclear activities of either party."

Agreement calls up on the parties that it "shall be implemented in a manner so that not to hinder or otherwise interfere with any other activities involving the use of nuclear material, non-nuclear material, equipment, components, information or technology and military nuclear facilities produced, acquired or developed by them independent of this Agreement for their own purposes" (section 2.4).[57] This failure of the Agreement to leave India's nuclear weapons programme unchecked has created serious apprehensions among Pakistan's strategic community as Agreement could allow India to improve its nuclear forces and this could force other stakeholders of the region to recalculate their strategic imperatives and this could trigger an arms race in the region.[58] The Agreement when viewed from Pakistan's perspective provides incentives to India in a number of areas which India can exploit to improve its deterrence capability but such a move could prompt other actors of the region to provide a similar response thus destabilising the entire South Asian region. In the first place, Agreement gives India leverage to segregate its nuclear programme into military and civilian ones and to keep a significant number of facilities outside the safeguards to serve its weapons programme. Under the "separation plan", India has pledged to place fourteen reactors under the safeguards but rest of the reactors (eight) will be outside the safeguards. Moreover, India has also managed to keep its Prototype Fast Breeder Reactor outside the safeguards net. India is free to use these reactors to improve its stock of fissile material for the production of nuclear weapons. According to Adil Muhammad Sultan, a Pakistani analyst the "separation plan" would enable India to improve its production of nuclear weapons from 6 to 10 to several dozen per year.[59] However, an increase in India's fissile material production could impinge the security interests of other states of the region, notably Pakistan and China and this could destabilise the region as was underlined by Robert J. Einhorn in his testimony before the Senate Foreign Relations Committee:

> *"If India steps up production [of fissile material], Pakistan can be expected to follow suit, China could decide to resume production, and others may be encouraged to seek their own production capabilities. The more materials produced, the more difficult and costly it will be to secure them, and the greater the risks of nuclear terrorism . . . stepped up Indian Production of fissile materials . . . could lead to increased tensions and destabilizing arms competition in southern Asia, involving India, Pakistan and China."*[60]

Secondly, Agreement has lifted decades old global moratorium on the supply of nuclear material and equipment to India. India can now easily access uranium from international market to feed its civilian nuclear programme. This will free India's domestic reserves of uranium which can be used to expand India's nuclear weapons programme. This has been affirmed by a Pakistani commentator Pervez Hoodbhoy who claims that "India will be freely able to import uranium for its safeguarded civilian reactors. This will free up the remainder of its scarce uranium resources for making plutonium. Further, when India's thorium fuelled breeder reactors are fully operational, India will be able to produce more bombs in one year than in last 30."[61]

Thirdly, Agreement leaves India's existing stockpiles of fissile material unattended by not urging India to place them under safeguards and does not place a ceiling on the future production of such materials. India already has a stock of more than 500 kilograms of weapons grade plutonium sufficient to produce more than 100 nuclear weapons. In addition to this it has 11.5 tons of reactor grade plutonium[62] in the form of spent fuel which can also be utilised to make additions to its already existing stockpile of nuclear weapons.

If India exploits these incentives, it can make both qualitative and quantative improvements in its nuclear weapons arsenal but such a move could also force Pakistan to strengthen its deterrence capability vis-à-vis India in view of the fact that Pakistan's strategic and nuclear build ups have always been India centric. This was affirmed by the former Pakistan's President Pervez Musharaf in his address to Foreign Correspondents Association of Philippines:

> *"Pakistan pursues the strategy of credible deterrence in both conventional and unconventional fields in accordance with the threat it may perceive. Pakistan...has quantified the strategy of minimum defensive deterrence and is refining its deterrence level of force...whenever an imbalance is created in the region; Pakistan has to balance it out in accordance with its strategy of minimum deterrence."*[63]

THE CHINA FACTOR

One of the driving forces behind the desire of the Bush administration to sign a nuclear pact with India was to balance the rising power of China by forging a close strategic relationship with India. China's growing economic and military strength is being perceived as a great strategic challenge by US policy makers. To face this challenge, India is being viewed as a key strategic partner.[64] The growing economic and political clout of

China is not only a cause of concern for US but also is being perceived as a great challenge in India. In 2002, the Planning Commission of India warned that "increasing economic and military strength of China may pose a serious challenge to India's security unless adequate measures are taken to fortify our [India's] strengths."[65] India's strategic analysts also consider China as a perpetual threat because of its close military and nuclear links with Pakistan and are in favour of India playing a proactive role in balancing the rising power of China.[66] So, balancing the growing economic and military power of China is an issue on which the interests of India and US converge and to face this challenge US has to transform India, if not a global power but at least a regional power so that it could act as a counterweight to China. Suggestions have been made by US based think tanks and strategic experts that US should provide every possible support to India to build its economic, defense and deterrence potential to prevent the Chinese dominance of the region and to further US strategic interests in the region. A report prepared by Ashley J. Telis of the Carnegie Endowment for International Peace provides a fascinating glimpse of the situation:

> *"If United States is serious about advancing its geopolitical objectives in Asia, it would almost by definition help New Delhi to develop its strategic potentials such that India's nuclear weaponry and associated delivery systems could deter against the growing and utterly more capable nuclear forces Beijing is likely to possess by 2025."*[67]

If "counterbalancing" China is one of the objectives behind the India-US strategic entente, then US has to provide enough leverage to India to develop a credible deterrence against China but such a move could have far reaching consequences for the region because growth of India's strategic nuclear deterrent could prompt China to respond programmatically by not only reviving its deterrence capability vis-à-vis India but it will also provide every possible support to Pakistan to build it as a counterweight against India. In such a scenario one could foresee a "triangular nuclear arms syndrome" in the region involving China, India and Pakistan.[68]

SINO-PAKISTAN NUCLEAR DEAL: A RESPONSE TO INDIA-US NUCLEAR DEAL

When India-US Civil Nuclear Cooperation Initiative was negotiated, a number of non-proliferation advocates expressed their deep reservations over the Agreement on the grounds that the Agreement would set a precedent for other states to cut similar deals with their allies and same came true when in 2010 China announced that it will build two nuclear reactors in Pakistan at Chashma nuclear site. Three years later, in 2013, it clinched another nuclear deal with Pakistan for the construction of two 1100 MW reactors at Karachi. The dollar 2.375 billion Chashma deal aims to set up two 650 MW reactors at Chashma site which already houses two reactors (Chashma-1 and Chashma-2) built by China. The work on the construction of new reactors (Chashma-3 and Chashma-4) has already begun and the reactors are expected to come in line by 2016 and 2017 respectively. Under the deal announced in 2013, China will supply two indigenous reactors to Pakistan each with a capacity of 1100 MWs. These reactor sales have triggered unease among the non-proliferation advocates in general and in India and US particular. The non-proliferation community considers the supply of new reactors to Pakistan as a violation of NSG norms and the strategic experts of India and US view it as a response to India-US nuclear deal. However, these concerns have been downplayed by China, arguing that the nuclear cooperation with Pakistan has a peaceful nature and is a response to acute energy crises of Pakistan. This was acknowledged by China's Foreign Ministry spokesperson, Hong Lei:

> *"I want to point out that relevant cooperation between China and Pakistan does not violate relevant norms of the NSG. In recent years, China and Pakistan have carried out some cooperation in the field of civilian nuclear cooperation. All this cooperation is for peaceful use and this cooperation is in compliance with our respective international obligations and subject to the safeguards of IAEA."*[69]

China became a member of NSG in 2004. Under the guidelines of cartel, it cannot supply nuclear material and equipment to a country that has not signed NPT and has not accepted full scope safeguards. Since Pakistan is neither a signatory of NPT nor has accepted full scope safeguards, so the supply of new reactors to Pakistan stands in contravention to the norms of the cartel. China, however, argues that the rector sales to Pakistan are a part of an agreement that predates China's entry into the cartel, therefore, the reactor sales have been 'grandfathered'. Under the 'grandfather' clause of the NSG guidelines, a country being a member of NSG can supply nuclear material and equipment to country not having a full scope safeguards in place if such supplies are covered by an agreement predating the entry of the supplier country into NSG.[70] Since China became a member of NSG in 2004 and argues that its recent reactor sales to Pakistan are part of an agreement signed with Pakistan in 1986, so the supply of reactors does not constitute a violation of NSG norms. However, NSG countries and non-proliferation experts do not buy the Chinese argument that sales have been grandfathered.[71] Because when China joined the NSG in 2004, it did not inform the group about its commitment to build additional reactors in Pakistan. It only declared to the NSG that under a longstanding agreement with Pakistan it is committed to "provide a second reactor, Chashma-2, more research reactors, plus supply of all the fuel in perpetuity for these units."[72] Fred McGoldrick, a non-proliferation expert, succinctly summarises the issue in following words:

> *"China now is citing the grandfather clause as justification for its planned assistance to Pakistan's civilian nuclear reactors. In a formal "declaration of existing projects" made at the time it joined the NSG in 2004, Beijing informed to NSG of its 1991 cooperation agreement with Pakistan under which it had supplied a 300-megawatt reactor at Chashma and had just undertaken to supply an additional 325 megawatt reactor at the same location. The Chinese claimed that the supply of these reactors did not constitute a "new" supply commitment and therefore was grandfathered under the NSG comprehensive safeguards guideline. This was a dubious claim at the time because the 1991 Chinese-Pakistani pact was a general framework agreement and reportedly did not contain an actual commitment to supply reactors at Chashma. In any Event, Beijing's 2004 explanations to the NSG did not mention grandfathering any more rectors under the 1991 agreement. Thus, the proposed supply of additional nuclear plants at the Chashma site would not be covered by the 1991*

agreement, would not be grandfathered under the NSG guidelines, and would clearly violate the commitments China made when joining the NSG in 2004..."[73]

Many non-proliferation experts blame India-US Nuclear Deal for the China's reactor sales to Pakistan. Henery Sokolski says that, "Two wrongs make a wrong, but it was to be expected once we [US] made the case for an exemption [for the US-India deal]."[74] Michel Krepon of the Henery L. Stimson Center says, "You can't help but hear about China supplying Pakistan with nuclear power plants and see it as a reaction to the US-India deal."[75] And, there are many who allege that by signing the nuclear deal with India, US has lost its leverage to protest against Sino-Pakistan nuclear deals. US knows that Pakistan has an appalling non-proliferation record. Growing instability and rise of extremist groups have further exacerbated the risk of Pakistan's nuclear assets falling into the hands of terrorist groups but it faces a dilemma how to oppose the deals when it itself has flouted the non-proliferation norms for geo-strategic gains.

NUCLEAR ENERGY AND INDIA'S ENERGY SECURITY

India is one of the fast growing economies of the world. In terms of primary energy consumption it is the fourth largest economy in the world after China, US and Russia and is likely to occupy third position by 2030. India's primary energy consumption has considerably increased during the past two decades but its per capita energy consumption is still much lower than the developed countries of the world. Country's per capita energy consumption is only 0.6 tonnes of oil equivalent (TOE) where as in Canada it is 7.6 TOE and in US it is 7.3 TOE.[76] India's economy has experienced unprecedented growth over the past several years. Between 2006 and

2010 Gross Domestic Product (GDP) of the country increased at a CAGR (Cumulative Average Growth Rate) of 8.2 per cent and during the same period country's primary energy consumption increased at a CAGR of 8.3 per cent [77] signalling a strong correlation between increased GDP and energy consumption. Currently India is the 9th largest economy in the world but is likely to become third largest by 2030 with a GDP of $ 15 trillion after China with a GDP of $53.8 trillion and US with a GDP of 38.5 trillion, according to Standard Charter's Super Cycle Report.[78] Since there exists a strong correlation between growth in GDP and energy consumption, higher rate of economic growth will also demand a higher rate of increase in country's energy production. According to India's Integrated Energy policy (IEP), to sustain a growth rate of 8 per cent India has to increase its primary energy supply three to four times and electricity supply five to six times over 2003-2004 levels.[79]

India's population is also increasing at an alarming rate and has already crossed 1 billion mark but most of the people still lack access to electricity. About 57 per cent of the rural households and 12 per cent of urban households of the country do not have access to electricity[80] and the areas which have been connected to grid, especially rural ones do not get reliable and continuous supply and often face blackouts. India also faces a challenge to provide clean and an affordable source of energy to a largest chunk of population living in the rural areas. Census 2011 reveals that nearly 80 per cent of rural households still use traditional sources of fuel like biomass for cooking purposes.[81] So providing access to clean, affordable and reliable sources of energy to large sections of population and exponential growth in population are the factors which will escalate the demand for energy in India and will put enormous pressure on fast depleting sources of energy.

Indian energy scenario is greatly dominated by the use of fossil fuels. In 2011 coal accounted for 52.9 per cent of India's primary energy consumption, where as oil and gas accounted for 38.8 per cent.[82] However, the domestic production of these sources has not been enough to keep pace with growing demand resulting in an increasing gap between demand and supply. To maintain a balance, India largely depends on fossil fuel imports. Since the prices of these fuels remain unstable and keep on changing thus making Indian economy vulnerable to volatile fuel prices. So, there is a need to switch over to an alternative source of energy ensuring a reliable supply of energy over a long period.

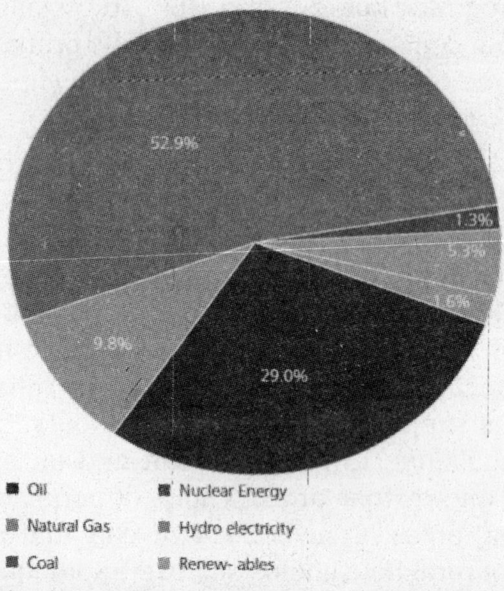

Fig. 4.2. India's Primary Energy Mix (2011)

Source: Securing Tomorrow's Energy Today: Policy and Regulations. Available at http://www. deloitte.com/assets/Dcom-India/Local%20Assets/Documents/IEC%202013/LongTerm Energy_Security.pdf.

Original source: BP Statistical Review 2012.

MAIN SOURCES OF ENERGY

Coal

Coal is the mainstay of India's energy mix accounting for more than 50 percent of India's primary energy consumption. According to India's Ministry of Power, of the total power generated in India as on 30-9-2013 coal accounted for 134388.39 MW which is 58.75 per cent of the total installed capacity. India is fourth largest producer of coal in the world after China, US and Australia, however, in terms of consumption it holds a third rank after China and US. India has fifth largest reserves of coal in the world. As on 1-4-2012 country's estimated reserves of coal stood at 2,93,497 Million Tonnes (MT) (Geological survey of India). Of the total estimated reserves, 33688.73 MT are coking coal, 258315.78 MT are non-coking reserves and rest are tertiary coal reserves.[83] Most of the coal deposits of the country are located in eastern parts of the country. Chhattisgarh, Jharkhand and Orissa together account for nearly 70 per cent of country's coal reserves. Andhra Pradesh, Madhya Pradesh and West Bengal account for about 26 per cent of the estimated reserves.

Table 4.2. Estimated Coal reserves (1-4-2012)

State	Reserves in Million Tonnes
Andhra Pradesh	22154.86
Assam	2.79
Bihar	160.00
Chhattisgarh	50846.15
Jharkhand	80356.20
Madhya Pradesh	24376.26
Maharashtra	10882.09
Orissa	71447.41
Sikkim	101.23
Uttar Pradesh	1061.80
West Bengal	30615.72
Total	**292004.51**

Source: Ministry of Coal

Original Source: Geological Survey of India

Note: The list does not include tertiary coal reserves which have been estimated to the tune of 1492.64 MT

Though country has fifth largest reserves of coal in world but domestic production has not been able to keep pace with growing demand making the country to import huge quantities of coal every year from the countries like Australia, Indonesia and South Africa. In 2010-11 India imported about 82 billion tonnes of coal and this figure is likely to go up in future and experts believe that by 3032 India will be importing coal worth 240-250 billion dollars.[84] The period between 2006-07 to 2011-12 experienced 7.40 per cent increase in country's coal consumption, however, during the same period production increased by 5.16 per cent only.[85] One of the reasons responsible for the poor performance of country's coal sector is its domination by public sector companies. In 2011- 2012, public sector accounted for more than 90 per cent of the total production dominated by Coal India Limited (CIL) where as private sector accounted for 9.1 per cent only.[86] Power sector is the largest consumer of coal in the country followed by steel and cement sectors. With the growth in power sector the demand for coal is likely to increase in coming years. According to the estimates of Central Electricity Authority (CEA) the demand for power sector by 2016-2017 is likely to reach to 913.22 MT.[87]

No doubt coal is a main source of India's energy production and is likely to remain a dominant player for at least next two decades but large scale use of coal has its own down sides. Combustion of coal leads to the emission of greenhouse gases with serious implications for environment. Moreover, Indian coal is characterised by high ash content, if not disposed of properly it can also cause great damage to human health and environment.

Table 4.3. Coal demand for power sector (Central Electricity Authority)

Year	In MT
2012-2013	637.21
2013-214	729.38
2014-2015	829.73
2015-2016	877.44
2016-2017	913.22

Source: Securing Tomorrow's Energy Today: Policy and Regulations. Available at http://www. deloitte.com/assets/Dcom-India/Local%20Assets/Documents/IEC%202013/LongTerm Energy_Security.pdf.

Oil and Gas

Oil and gas constitute an important component of India's primary energy mix accounting for 39 per cent of country's primary energy mix in the year 2011. In 2011, India was the fourth largest importer and consumer of oil and petroleum products in the world after US, China and Japan. As on 31-3-2012 country had 759.59 MT of estimated reserves of crude oil, mostly located offshore. Western offshore accounts for about 45 per cent of the estimated reserves followed by Assam which accounts for about 23 per cent.[88] Country holds very limited reserves of oil just 0.7 per cent of world's proven reserves. But accounts for 3.9 per cent of global consumption making the country largely dependent on imports. Imports account for about 73 per cent of India's oil consumption.[89]

India had 1330.26 billion cubic metres (BCM) of estimated reserves of natural gas as on 31-3-2013, mostly located eastern offshore (34.73 per cent). Like oil reserves, India has very limited reserves of gas just 0.8 per cent of world's proven reserves but accounts for 1.9 per cent of global gas consumption making country to import huge quantities of gas, nearly 20 per cent of its consumption to satisfy the demand.[90] India first time began to import natural gas from Qatar in 2004 when consumption outstripped domestic production. In 2011 India was the 6th largest importer of natural gas in world.

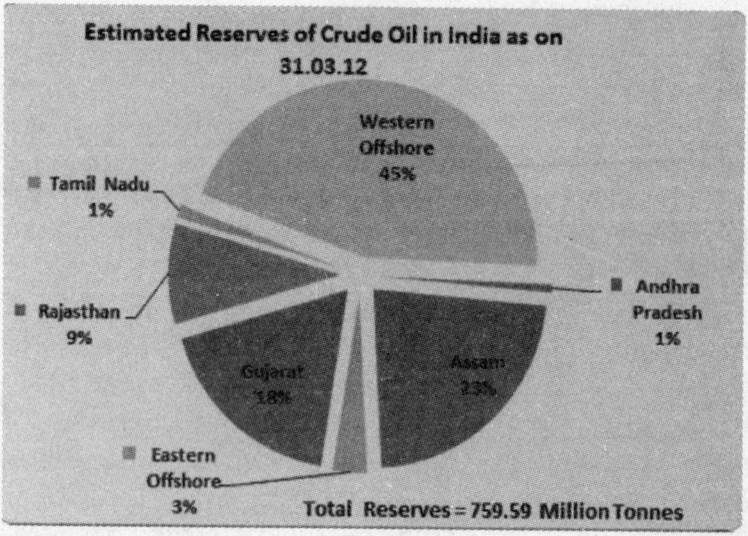

Fig. 4.3

Source: Energy Statistics 2013, Report of Ministry of Statistics and Programme Implementation

Since 2001, gas consumption in India has annually grown at 10 per cent rate mainly driven by the growth in power and fertilizer sectors. In 2010 power sector accounted for 45 per cent of total gas demand and fertilizer sector accounted for 28 per cent.[91] According to the estimates of International Energy Agency (IEA) gas demand in India is likely to reach to 94 billion cubic metres (BCM) by 2020 and 132 BCM by 2030 mainly driven by the growth in power and industrial sectors.[92] Keeping in view the limited reserves of gas and increasing demand Government of India is seriously pursuing the options to secure its gas imports in future. One of these is acquisition of foreign gas assets and second is building a strong pipeline infrastructure through international projects to import gas from Central Asian countries. One such project is Turkmenistan-Afghanistan-Pakistan-India (TAPI) project. The negotiations on the project have been going on from last one decade and considerable progress has been made. Once realised, the project will go a long way in satisfying India's increasing demand for gas.

Hydro

India is blessed with a vast hydropower potential to the tune of 148700 MW as per the estimates of Central Electricity Authority, however, only 26 per cent of this potential has been tapped so far. India ranks fifth in world in terms of exploitable hydropower potential. Of the estimated potential more than 86 per cent lies in North, Southern and North-Eastern regions of the country and the rest in Western and Eastern regions of the country.

Table 4.4. Region wise hydropower assessed potential

Region	Potential in MW
North	53350
Western	10050
Southern	19045
Eastern	9890
North Eastern	56365
Total	**148700**

Source:www.worldenergy.org/documents/congresspapers/376.pdf

Despite being a cheap, clean and a preferred source of energy, the exploitation of this source has not been up to a desired level due to a number of factors including environmental concerns, land acquisition problem, long gestation period, rehabilitation concerns, and long clearance and approval procedures. It is because of these constraints that share of hydropower in India's power generating capacity has been considerably falling. In 1963 hydropower accounted for 50 per cent of country's total installed capacity, however, in 2010 its share fell to 25 per cent[93] and currently it accounts for 17 per cent of the total installed capacity. Under the 11th five year plan the target was to add 15627 MW but only 35 per cent of this target was achieved. The 12th plan aims at a capacity addition of 10897 MW which is 12 per cent of the total addition envisaged, however, keeping in view the challenges confronted by the sector analysts claim that that the target may fall short by 30 per cent.

Two serious issues confronting the sector are environmental concerns and social displacement. Though hydropower generation does not emit greenhouse gases but large dams lead to submerging of vast area of land and forests thus posing a great threat to environment and local communities. As per estimates about 40-80 million people have been displaced because of large hydroelectric dams. In India and China about 25-58 million people have been displaced since 1950 to 1990 and about 70 per cent of the displaced people in India have not been rehabilitated yet.[94] Due to widespread protest by environmental activists and displaced sections of people, a number of projects in India have come to standstill. For instance 400 MW Maheshwari Project which was started in 1990s is still incomplete because of stiff resistance posed by Narmada Bachao Andolan. Same is case with 2000MW Lower Subansiri Hydropower Project in Aruncahal Pradesh. There are at least 26 hydropower projects with an installed capacity of 1924MW which have been approved by CEA but are still to get clearance from Environment Ministry.[95]

Table 4.5. Basin wise hydropower assessed Potential

Name of Basin/ River system	Potential in MW
Indus Basin	33832
Ganga Basin	20711
Central Indian River System	4152
West Flowing Rivers of Southern India	9430
East Flowing Rivers of Southern India	14511
Brahmaputra Basin	66065
Total	**148701**

Source: www.eai.in/ref/ae/hyd/hyd/html.

Nuclear

Nuclear power is the fourth largest source of electricity in India after thermal, hydro and renewable sources accounting for 2.08 per cent of the total installed capacity. Currently, India has twenty-one nuclear reactors in operation with a capacity of 5.3 GW and has six projects under construction with a capacity of 4.3 GW. Though nuclear power has a marginal presence in country's energy mix but the country expects to generate 20000MW of nuclear power by 2020 and about 63000 MW by 2032. These ambitious plans have been largely motivated by the signing of Civil Nuclear Cooperation Agreement with US. The signing of the Agreement in 2008 has ended India's decades long nuclear isolation. The Agreement enables India to carry out trade in nuclear fuel and technology with advanced countries of the world and provides it access to latest nuclear technology and fuel which has been a main constraint of its nuclear energy programme. Since the signing of the Agreement in 2008, India has inked civil nuclear cooperation agreements with a number of countries including Canada, France, Russia, Kazakhstan etc. However, the achievement of targets set forth will also depend to a greater extent on the successful implementation of India's three stage nuclear programme which aims to exploit vast thorium reserves of the country. India has very limited reserves of uranium about 50,000-70,000 tonnes but is endowed with vast reserves of thorium about 650,000 tonnes. To tap this resource

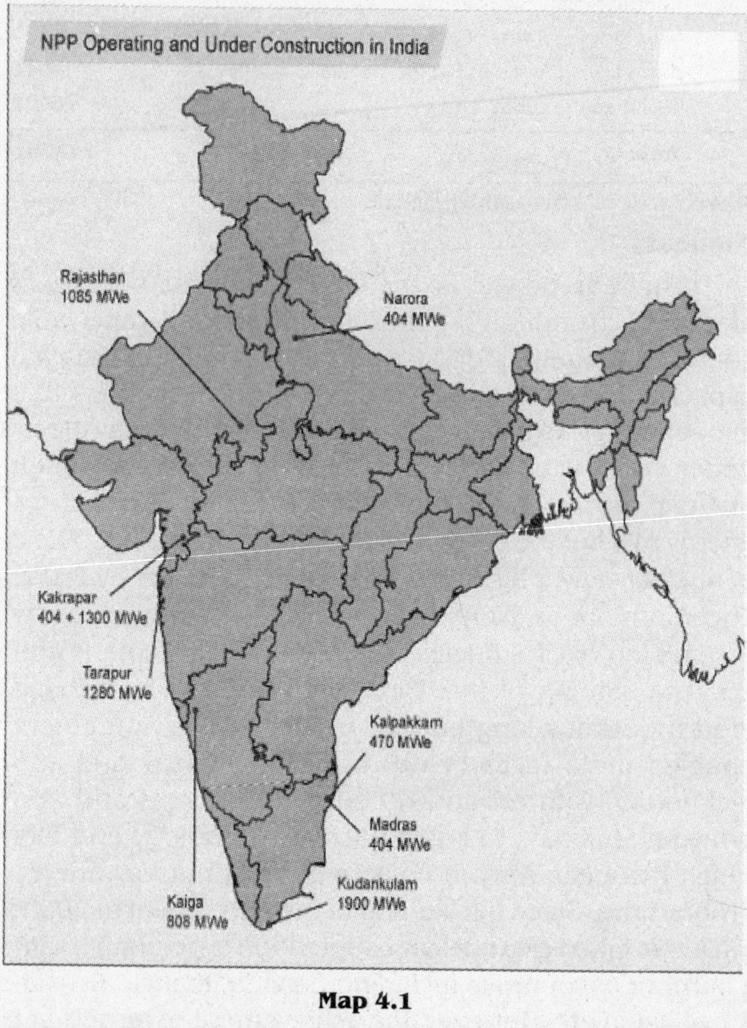

Map 4.1

Source **:** http://www.world-nuclear.org/info/Country-Profiles/Countries-G-N/India/

country has embarked on a three stage nuclear programme. The first stage consists of PHWRs and LWRs which produce plutonium as a by-product which will be used as fuel along with thorium in second stage which consists of breeder reactors. The breeder reactors besides producing more plutonium will also convert thurium-232 in to uranium-233. The third stage is based on Advanced Heavy Water Reactors. These reactors will burn Uranium-233 produced in the second stage along with thorium-232 converting it into uranium-233 thus producing more and more fuel. However, only the first stage of this programme has been implemented so far.

Renewable Sources

Renewable energy constitutes a vital part of India's energy mix and is increasingly becoming an attractive source of energy not only in India but across the globe. Increasing emission of greenhouse gases, rising prices of fossil fuels and increasing competition of their procurement are some of the factors which have led to a renewed interest in the renewable sources of energy among the nations. In India renewables excluding large hydro account for 12.32 per cent of the total installed capacity. If large hydropower capacity is also taken into account then the share will rise to 29.71 per cent of the total installed capacity. The share of renewable energy in India's energy mix has been considerably rising over the years mainly because of favourable government policies and conducive environment. The share of renewable energy in India's energy mix increased from 7.8 per cent in 2008 to 12.3 per cent in 2013. Under the 12th five year plan the target is to add 29.8 GW of renewable energy taking the capacity to 55 GW. The major share is likely to come from wind (15 GW), followed by solar (10 GW), biomass (2.7GW) and small hydro (2.1 GW).[96] Greenpeace International and European Renewable Energy Council

(EREC) have estimated that by 2050 about 69 per cent of India's electricity will come from renewables.[97] For the promotion of renewable energy, country has set up separate a ministry called Ministry of New and Renewable Energy (MNRE). India is the only country having an exclusive ministry dedicated to the development of renewable energy. The main sources of renewable energy in India are wind, small hydro, solar and biomass.

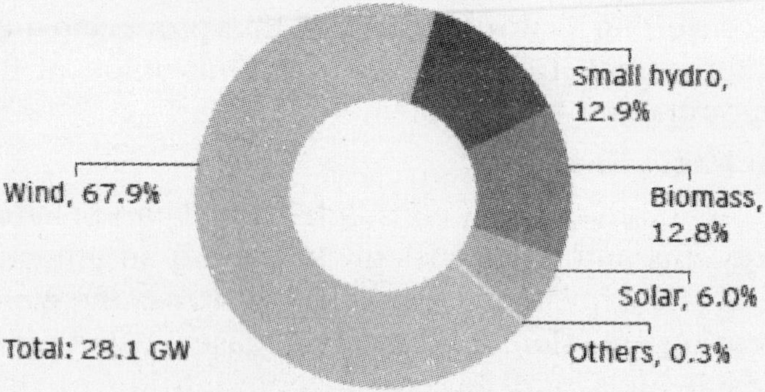

Fig. 4.4. India's installed capacity of renewable energy as on 31 March 2013.

Source: Mapping India's Renewable Energy Growth Potential: Status and Outlook 2013. Available at www.ey.com/Publication/vwLUAssets/Mapping_Indias_Renewable_Energy_growth_potential/$FILE/EY-Mapping-India's-Renewable-Energy-growth-potential.pdf.

Original Source: Ministry of New and Renewable Energy

Wind : Wind is a clean, safe and environmentally benign source of energy. Low gestation period and low maintenance and operating costs further make it an attractive source of energy. Wind is an abundant source of energy and world has enough wind potential to meet the power requirements of the people from this source. According to a study conducted by Christina Archer and Mark Jacobson of Stanford University, world has a wind power potential of 72 terawatts and tapping just 20 per cent of this potential can satisfy the needs of all the countries of the world.[98] In India, renewable energy profile is dominated by wind accounting for 67. 9 per cent of total installed renewable capacity as on 31-3- 2013. India's wind power potential has been estimated to the tune of 49130 MW.[99] As of 31- 12- 2013 about 20149 MW of this potential had been tapped (MNRE). Most of the installed capacity is concentrated in the states of Tamil Nadu, Maharashtra, Gujarat, Karnataka and Rajasthan. Currently India is the fifth largest wind energy producing country in the world after US, China, Germany and Spain.

Small Hydro : In India hydropower projects having a capacity up to 25 MW are designated as small hydropower projects. India has an estimated potential of about 15000 MW from small hydro. Most of the potential lies in Himalayan states. Of the estimated potential about 3763 MW have been harnessed (as of 31-12-2013). As of 31 March 2013, a total of 967 small hydropower projects having a capacity of 3632 MW including 329 private sector projects aggregating a capacity of 1748 MW had been set up in different parts of the country. In addition to this there are about 327 projects in the country which are at different stages of development.[100]

Power from small hydropower projects is a cheap and environmental friendly source of energy. It emits least amount of greenhouse gases and does not lead to dislocation of large sections of population, a challenge associated with large hydropower projects. Small hydropower projects are the most suitable means to provide energy to remote and far-flung areas which are not connected to grid.

Solar Power : India is blessed with huge solar power potential but this potential is largely untapped. Country has an estimated solar power potential of about 50000 MW. As of 31-3-2013 solar energy accounted for 6 per cent of the total installed capacity from renewable sources.

By the end of March 2010 more than thirty-five grid connected solar connected photovoltaic plants aggregating a capacity of 1028 MW were installed in country.[101] For the promotion of solar energy government of India has launched Jawaharlal Nehru National Solar Mission in 2010. The mission aims to develop 20000 MW of grid connected solar power and 2000 MW of off-grid power by 2022. The mission has been instrumental in improving country's solar sector as country's installed solar power capacity increased from merely 30 MW to more than 2000MW in 2013.[102] Solar energy is clean source of energy and offers a great opportunity to India to reduce its emission of greenhouse gases. According to a recent World Bank report, 'Paving the Way for a Transformational Future: Lessons From JNNSM phase 1', solar power can help India to reduce its dependence on fossil fuel imports and will help it to cut its emission of greenhouse gases 20-25 percent over 2005 levels by 2020.[103] However, it is an expensive source of energy when compared with the power generated from coal and gas fired plants which cost about RS 4.5 per kilowatt-hour where as solar power costs about Rs 7 per kilowatt-hour. However, prices have sharply come down from past several years from Rs 18 per kilowatt-hour in 2010 and with further expansion of the sector the prices are likely to come down to rupees 5.0-5.5 a kilowatt hour.[104]

Biomass : Biomass is an important and an old source of energy. In India, most of the people particularly living in rural areas mainly depend on biomass for cooking, heating and lighting purposes. Census 2011 reveals that 62 per cent of rural households use firewood, 12.3 per cent use crop residue and 10.9 per cent use dung as primary fuel for cooking. In contrast to this only 20 per cent of urban households use firewood as cooking fuel and just 2 percent use dung or crop residue.[105] About

70 per cent of the population still depends on biomass as primary energy fuel. In India, huge quantity of biomass is available in the form of husk, straw, bagasse, jute, cotton etc. and a significant quantity is used in power sector. Country has an estimated potential of about 22,000 MW from biomass. Most of this potential is concentrated in the states of Punjab, Maharashtra, Uttar Pradesh, Tamil Nadu, Karnataka and Madhya Pradesh. As of 31 March 2013 country had an installed capacity of 3601 MW of power from biomass.[106] Country has about 800-900 biomass power plants making a significant contribution to country's renewable power capacity.

NUCLEAR ENERGY: CAN IT BE A FUTURE SOURCE?

Energy is the basic requirement of every country. It is key to economic growth and sustainable development, therefore, securing access to vast and affordable sources of energy is the prime concern of every country. India is one of the fast growing economies of the world and is expected to become third largest economy by 2030. As Indian economy grows in size the demand for energy is likely to escalate putting tremendous stress on country's fast depleting energy sources. Currently country's energy requirements to a greater extent are satisfied by hydro carbons but for a long time perspective country needs a robust source of energy. But any source of energy having a potential to satisfy country's long term energy requirements must full fill certain criteria before it is considered as a viable option. It must be sustainable, cost competitive, safe, socially acceptable and over and above environmental friendly.

In India, where energy requirements are escalating at an alarming peace and environmental concerns are a gaining importance, nuclear energy is being portrayed as best solution to climate change crises and energy security concerns. The issue has been put in perspective by Indian Prime Minister in following words:

> *"Our long term economic growth is critically dependent on our ability to meet our energy requirements of the future... [Since] our proven reserves of coal, oil, gas and hydropower are totally insufficient to meet our requirements, (and) the energy we generate has to affordable, not only in terms of financial cost but in terms of cost to our environment; this is the reason why we place so much importance on nuclear energy."*[107]

For the large scale growth of nuclear industry Indian nuclear establishment has embarked on a large scale expansion of nuclear power programme. The plans are to generate 20000 MW of nuclear power by 2020, 63000 MW by 2032 and 275000 MW by 2050. However, keeping in view the past record of India's nuclear industry, growing concerns regarding the use of nuclear energy and its impact on environment and human health and increasing use of fossil fuels two questions need to be asked at outset: Can nuclear energy become India's future source of energy? And can it help to mitigate climate change problem by reducing the emission of greenhouse gases?

Department of Atomic Energy has set a target to generate 20000 MW of nuclear power by 2020. The achievement of this target is going to be herculean task for DAE as the target involves certain unrealistic projects:

- Completion of all ongoing projects including 500MW PFBR.
- Upgrading 500 MWe PHWRs to 700 MWe.
- Building eight additional 700 MWe PHWRS and four additional Fast Breeder Reactors.
- Importing six LWRs each with a capacity of 1000 MWe.[108]

It is not first time that DAE has come up with such ambitious targets. Such targets had been set in the past too but it miserably failed to achieve them. As early as 1962 Homi Bhaba, the founder of India's nuclear programme announced that country will produce 20000-25000 MW of nuclear power by 1987. These targets were revised by his successor Vikram Sarabhai who predicted that nuclear sector would add about 43000 MW to country's energy basket by 2000.[109] But none of these targets were realised. Today even after more than forty years of commercial operation, the industry has been able to generate 4780 MWe only, which is just 2 per cent of country's total installed capacity.

The targets set by DAE will also depend to a large extent on the procurement of uranium, which is used as fuel in PHWRs and on expanding the base of nuclear infrastructure in the country. Country has very limited reserves of uranium and there is a wide gap between demand and supply, which is largely responsible for the low capacity factor of India's nuclear power reactors. To bridge this gap there is an urgent need to scale up uranium supplies but nuclear establishment is facing a stiff resistance to open up new uranium mines or to set up new plants. A strong anti-nuclear movement is growing across the country. The protests against Kundankulam nuclear power projects and proposed Jaitapur projects on environmental and health reasons are some recent examples of growing anti-nuclear resistance in the country.

The achievement of the targets set by DAE will also be determined to a greater by the successful implementation of India's breeder programme. The first stage of nuclear programme which is based on PHWRs can support only 10000 MWe, so much will depend on plutonium fuelled breeder programme which is yet to take off commercially even after more than three decades of research and development. Moreover, breeder technology is a complex, expensive and unreliable technology. The huge costs involved in the construction of breeder reactors, difficulties in operation and safety concerns have compelled many countries to abandon their breeder projects. France is one of them which abandoned its Super Phinex breeder project in 1997 because of number of accidents and difficulties involved in operation.

Given the poor track record of India's nuclear industry and growing anti- nuclear resistance in the country, it seems highly unlikely that DAE will be able to realise the proposed targets and even if the targets are met nuclear power would account for a small amount of projected electricity capacity in 2020 (8-10 per cent) and about 20 per cent in 2052.[110] India's energy scenario is currently dominated by the fossil fuels and this trend is likely to continue for a longer period.

Nuclear energy is also being promoted as a solution to rising greenhouse gas emissions. However the scope of reducing the emission of greenhouse gases through the increased use of nuclear energy is very slim. This is possibly because of two reasons. First, Nuclear energy per se is not a clean source of energy. During the different stages of nuclear power generation‾ mining, milling, reactor construction‾ a large quantity of greenhouse gases is released into the atmosphere. Second, power sector accounts for a small fraction (only 9 per cent) of global green house gas emissions.[111] Other sectors of economy like transportation and industry account for a sizeable amount of greenhouse gas emissions and these sectors do not necessarily operate on electricity which nuclear energy is best suited to produce. In Japan, from 1966 to 1995, the share of nuclear power scaled up from zero to 40000 MW but during the same period its emission of carbon dioxide increased three times from 400 million tonnes to 1200 million tonnes[112] indicating that nuclear energy can do little to reduce the emission of greenhouse gases.

NUCLEAR ENERGY: IS IT SAFE?

Nuclear energy is a robust source of energy. One kilogram of uranium can generate as much electricity as 1500 tonnes of coal.[113] However, the great challenge associated with this source in terms of safety make it an unattractive source of energy. This is the only source of energy which can cause catastrophic accidents like Chernobyl and Fukushima. Studies have revealed that every stage of nuclear fuel cycle from mining to decommissioning carries with it immense risks to human health and environment. The adverse effects of nuclear fuel cycle can be first seen among uranium miners because of their exposure to lethal radioactive substances like radium-226 and radan-22. Radium-226 if swallowed can cause fatal diseases like bone cancer and leukaemia. Radon-22 is a radioactive gas; its inhalation can cause lung cancer. It has been found that one-fifth to one-half of uranium miners in North America have lost their lives because of lung cancer.[114] Same is the case with the people living in the vicinity of a nuclear facility. These people are more vulnerable to severe health risks like congenital deformities, solid tumours and other fatal diseases. According to a study conducted by the researchers at the Medical University of Carolina, USA, the people living near to nuclear facilities, particularly children and young people have increased rates of leukaemia.[115] The greatest danger, however, emanates from a nuclear accident which could happen during the routine operation of a nuclear facility and such an accident could have catastrophic consequences as revealed by Chernobyl nuclear accident which continues to haunt human memories even after more than two decades. Though the actual damage caused by the accident is not yet known and probably can never be known, but some recent studies have come up with astonishing figures. According to a recent study based on over 5000 scientific papers and radiological surveys carried out by Russian biologist Alexei Yablokov and two researchers from Ukraine, till 2004 the disaster has claimed about 985000 lives. Ukraine Health Ministry however estimates that about 530000 people have died till 2004.[116] Though these figures have been contested by IAEA and World Health Organisation (WHO) but the fact remains that Chernobyl nuclear disaster has caused an irreparable damage to human health and environment. A recent study of American Cancer Institute has revealed that the children who were exposed to radioactive material because of the accident still continue to develop thyroid cancer [117] and radioactive elements from the disaster site are still leaking, causing a severe damage to environment.

After the Chernobyl nuclear disaster in 1986, nuclear industry did not experience any major nuclear accident till 2011 and votaries of nuclear energy were expecting a nuclear renaissance but nuclear industry was shattered when in 2011 Japan's Fukushima nuclear plant lost its cooling system resulting in one of the world's nuclear disasters. The accident brought to the light that generating electricity through nuclear means is a dangerous business and accidents are inevitable with this complex technology irrespective of safety measures. Such was the impact of the accident that a number of countries including Germany, Sweden, Italy and Japan decided to phase out their nuclear power programmes. The tragedy forced the Japan to turn away from nuclear power and switch over to renewable energy. The country has decided to close down its all operational nuclear reactors by 2040. The deficit in energy is likely to be compensated by the large scale use of renewables. France, where more than fifty per cent of electricity is nuclear generated is even considering to reduce its heavy dependence on nuclear energy. So, before going for a major expansion of its nuclear power programme Indian nuclear establishment must think that can a densely populated India afford a nuclear accident like Chernobyl or Fukushima.

NUCLEAR ENERGY: IS IT CHEAP?

Another unattractive feature of nuclear energy is that it is an expensive source of electricity. Way back in 1950, C.G. Suits, Vice President and Director of research at General Electric told the American Association for the Advancement of Science that:

> *"At present, atomic power presents an exceptionally costly and inconvenient means of obtaining energy which can be extracted more economically from conventional fuels... The economics of atomic power are not attractive at present, nor are they likely to be for a long time in the future. This is expensive power, not cheap power as the public has been led to believe."*[118]

The statement made in 1950 when nuclear industry was in embryonic stage continues to have relevance even today, after more than five decades of research and development in the nuclear field. A number of studies have shown that nuclear energy costs more than the energy generated from coal and gas fired plants. A 2003 study of Massachusetts Institute of Technology(MIT) updated in 2009 concludes that the "'levelised' cost of electricity generated by a new nuclear power plant is about

30-35 per cent higher than the cost of electricity from a coal fired or combined cycle gas turbine plant."[119] A comparative study of Kaiga Atomic Power Stations and Raichur thermal power stations in India carried out by M.V. Ramana also reveals that nuclear power is more expensive than the power generated by coal power plants.[120] Nuclear reactors are capital intensive dominated by huge construction and repairing costs. Hefty amount is also required to dispose off the nuclear waste generated by nuclear plants and to decommission a nuclear facility after it operating life is over. The costs associated with this process are so high that US Energy Consultant Amory Lovins says that, "Today, if a nuclear power plant costs nothing to build, it would be cheaper to write it off... than to operate."[121] In US, number of nuclear reactors (2-5) will be shut down in next five years possibly because of the out dated equipments. The industry does not find it economical to replace the out-dated equipments with new and modern ones.[122]

NUCLEAR WASTE: AN EVERLASTING THREAT

A major concern associated with nuclear power generation is the safe disposal of nuclear waste generated by nuclear power plants which remains radioactive for millennia, thus posing a great threat to health and environment. For instance the half life of uranium 235 is 710 million years and that of uranium 238 is 4.5 billion years. Bittu Sahgal, editor of the *Sanctuary Magazine* sets out the issue quite clearly in terms of the damage caused by nuclear waste to the environment and costs involved in mitigating such damage:

> *"If the centralised bureaucracy of Mauriya Kings two thousand years ago had discovered nuclear power, we in India and Pakistan would probably be spending half of our*

national budget storing and caring for or repairing the damage done by atomic waste."[123]

An average nuclear reactor annually produces about 20-30 tonnes of highly radioactive waste. This waste is kept in temporary storage sites and no country has yet found a way to dispose it off safely. An attempt was made by US to store its nuclear waste in a permanent repository in Yucca Mountains but the project ran in to trouble leading to its termination in 2010. Obama administration has constituted a commission called Ribbon Commission to find a permanent solution to the nuclear waste.[124]

In many countries of the world the waste from nuclear reactors is reprocessed to minimize its quantity but reprocessing is not a permanent solution. Reprocessing itself leads to the generation of huge quantities of radioactive waste which is released into the oceans or atmosphere. La Hague reprocessing facility in France is the largest man made source of radioactive waste. The facility annually discharges 370 million litres of radioactive waste into English Channel.[125] Other reprocessing facilities in U.K., India and elsewhere are also significant sources of radioactive waste.

Table 4.6. Life span of radioactive materials

Radioactive material	Half life span
Iodine 131	8 days
Ruthenium 103	39 days
Ruthenium 106	One year
Strontium 90	30 years
Caesium 134	Two years
Caesium 137	30 years
Plutonium 239	24, 100 years
Uranium 234	247,000 years
Uranium 235	710m years
Uranium 238	4.5bn years

Source: The Hindu, 25 April, 2011.

CONCLUSION

A thorough analysis of India-US Civil Nuclear Cooperation Agreement suggests that the Agreement has the potential to fuel India's weapons programme, thus providing a great challenge to global non-proliferation regime on the one hand and to the stability of the South Asian region on the other hand. The Agreement does not place any restraints on India's fissile material production or on the production of nuclear weapons. The Agreement provides India access to surplus nuclear fuel which will free her domestic reserves of uranium which India can exploit to expand her weapons programme. However, an augmentation of India's nuclear forces can prompt other states of the region to respond which can trigger an arms race in the region. An analysis of India's energy profile also leads to the argument that the Agreement can play a marginal role in meeting India's rising demand of energy keeping in view the past performance of India's nuclear energy sector and challenges associated with the use and development of nuclear energy. India's energy scenario is currently dominated by fossil fuels and this trend is likely to continue for a longer period till another source of energy, probably renewables, is fully developed.

REFERENCES

1 Ashley J. Tellis, 2005, "Should US Sell Nuclear Technology to India?-Part-II," *Yale Global*, 10 November.

2 Quoted in Harsh V. Pant, 2011, *The US-India Nuclear Pact: Policy Process and Great Power Politics*, New Delhi: Oxford University Press, p. 116.

3 Quoted in Gerald Felix Warburg, November 5, 2012, "Non-Proliferation Policy Crossroads," *The Non-Proliferation Review*, vol. 19, no. 3, p. 462.

4 "Issues and Questions on July 18 Proposal for Nuclear cooperation with India." Available at http://www.armscontrl.org/pdf/2005/118_India_Ltr_congress.pdf, (accessed on 12 September 2013).

5 Zia Mian and M.V Ramana, "Wrong Ends, Means and Needs: Behind the US Nuclear Deal with India." Available at http://www.armscontrol.org/oct/2006/01-02/JANFEB-indiafeature#bio, (accessed on 1 September 2013).

6 *Ibid.*

7 Quoted in Daryl G. Kimbal, July 26, 2006, "Impact of the Nuclear Deal on India's Fissile Material Production Capacity." Available at http://www.armscontrrl.org/print/549, (accessed on 16 November 2013).

8 Mark Fitzpatrick, 2008, "US-India Nuclear Cooperation Accord: Implications for Non-Proliferation Regime," *Asia Pacific Review*, vol. 15, no.1, p. 80.

9 Strobe Talbot, 2005, "Good Day for India, Bad for Non-Proliferation," *Daily Times*, 25 July.

10 Quoted in Harsh V. Pant, *The US-India Nuclear Pact: Policy Process and Great Power Politics, op. cit.*, pp. 128-129.

11 Mark Fitzpatrick, "US-India Nuclear Cooperation Accord: Implications for Non-Proliferation Regime," *op. cit.*, p. 78.

12 Michael Krepon, September 2008, "Likely Consequences of the Nuclear Suppliers Group Decision."Available at http://www.stimson.or/essays/likely-consequences-of-the-nuclear-suppliers-group-decision/, (accessed on 1 November 2013).

13 Michael Krepon, September 1, 2008, "US-India Nuclear Deal Now Done, Reckoning to Come Later." Available at http://www.stimson.org/essays/us-india-nuclear-deal-now-done-reckoning-to-come-later, (accessed on 7 November 2013).

14 "Deal Will Trigger New Crises, says Iran," 2008, *The Hindu*, 6 October.

15 Harsh V. Pant, *The US-India Nuclear Pact: Policy Process and Great Power Politics, op. cit.*, p. 104.

16 *Ibid.*, p. 124.

17 Dr. Harsh V. Pant, "The US-India Nuclear Deal: The End Game Begins," January 29, 2006. Available at http://www.worldsecurityntework.com/India/Pant-Dr-Harsh-V/The-US-India-Nuclear-Deal-The-End-Game-Begins, (accessed on 27 September 2013).

18 Dr. Ahmad Rashid Malik, 2006, "Japan's Reaction to Indo-US N-Deal," *The Nation*, 6 March.

19 Leonard Weiss, November 2007, U.S. - India Nuclear

Cooperation," *The Non-Proliferation Review*, vol.14, no. 3. pp. 441-442.

20 Quoted in George Perkovich, "The US-India Nuclear Deal," pp. 3-4. Available at http://www.icnnd.org/Documents/US_India_Nuclear_Deal.doc, (accessed 10 November 2013).

21 Quoted in C. Raja Mohan, 2006, *impossible Allies: Nuclear India United States and the Global order*, New Delhi: Indian Research Press, pp. *194-195.*

22 Ashley J. Tellis, "Should US sell technology to India?- Part-II", *op. cit.*

23 Quoted in Gurmeet Kanwal, "Indo-US Nuclear Deal and Non-Proliferation: Some Views From US." Available at http// www.ipcs.org/article/nuclear/indo-us-nuclear-deal-and-non-proliferation-some-views-from-2368html, (accessed on 23 October 2013).

24 Daryl G. Kimbal, "Seeing through Spin: "Critics" Rebut White House on the US-India Nuclear Cooperation Plan." Available at http://www.armscontrol.org/pressroom/2006/20060309-india-cretics-rebut-WH, (accessed on 12 October 2013).

25 Dr. Harsh V. Pant , "The US-India Nuclear Deal: The End Game Begins," *op. cit.*

26 David Albright and Susan Basu, April 5, 2006, "Neither a Determined Proliferator, Nor a Responsible State: India's Record Needs Scrutiny." Available at: http://www.isis-online.org/uploads/isis-reports/documents/indiacritique.pdf, (accessed on 14 November 2013).

27 Quoted in Siddharth Varadarajan, 2005, "Nuclear Bargain May prove Costly in the Long Run," *The Hindu, 20* July.

28 "N Deal Will Hit Our Strategic Programme," 2007, Interview of Brijesh Mishra with Karan Thapar, *The Pioneer*, 9 August.

29 Bharat Karnad, 2005, "Remember the Tritium," *The Asian Age*, 7 August.

30 "Indian PM's Statement in Parliament: We Will Shut down CIRUS Reactor, Shift Apsara Reactor," March 7, 2006. Available at http://in.rediff.com/news/2006/mar/07ndeal2.html, (accessed on 23 November 2013).

31 A. Gopalakrishnan, August 27 - September 2, 2005, "Some Concerns on Indo-US Deal," *Economic and Political Weekly*, vol. 40, no. 35, p. 3803.

32 P.K. Iyengar and M. Gupta, 2006, "India-US Deal Will Destroy

Nuclear Research," *The Asian Age*, 5 April.

33 A. N. Prasad, 2006, "Down to the Last Atom," *Outlook*, 27 February.

34 P.K. Iyengar and M. Gupta, "India-US Deal will Destroy Nuclear Research," *op. cit.*

35 Zia Mian et al, September 2006, "Fissile Materials in South Asia: The Implications of US-India Nuclear Deal," A Research Report of International Panel on Fissile Materials, pp. 3-4.

36 *Ibid.*, p. 4.

37 *Ibid.*, p. 4 and 12.

38 Zia Mian and M.V. Ramana, August 27 - September 2, 2005, "Feeding the Nuclear Fire," *Economic and political weekly*, vol. 40, no. 35, p. 3811.

39 Zia Mian et al, "Fissile Materials in South Asia: The Implications of US-India Nuclear Deal," *op. cit.*, p. 24.

40 P. K. Iyengar, 2009, "The Indo-US 'Unclear Deal'" in P. K. Iyengar et al, Strategic *sell-Out: Indian-US Nuclear Deal*, New Delhi: Pentagon Press, pp. 109-110.

41 Bharat Karnad, 2006, "Why the Nuclear Deal is A Disaster," *The Asian Age*, 18 March.

42 Bharat Karnad, 2006, "Turning India into Nuclear Cripple," *The Asian Age*, 7 April.

43 Sonika Gupta et al, September 2006, "Assessing the Indo-US Nuclear Deal on Civil Nuclear Cooperation: Forging a New Partnership," Report of National Institute of Advanced Studies, p. 7.

44 See Bharat Karnad, "Heading into Nuclear Oblivion" in P.K. Iyengar et al, Strategic *sell-Out: Indian-US Nuclear Deal, op.cit.*, and Sandhya Jain, September 13, 2008, "Manmohan Outsources Political sovereignty," *Mainstream*, vol. XLVI, no. 39.

45 Bharat Karnad, 2006, "Parliament Must Examine PM's Nuclear Deal," *The Asian Age*, 14 May.

46 Cited in Arundhati Ghose, 2007, "123 Agreement: No Threat To India's Strategic Programme," *The Tribune*, 12 November.

47 Justine Isola, August 26, 2010, "If India Tests? The Implications for India-US Civil Nuclear Deal," *IDSA Issue Brief*, p. 5

48 *Ibid.*, p. 10.

49 Qudssia Akhlaque, 2006, "Islamabad not fully Informed on N-Deal," *Dawn*, 11 April.

50 Quoted in Syed Shahid Hussain Bukhari, January 2011, "India-United States Strategic Partnership: Implications for Pakistan," *Barkley journal of Social Sciences*, vol.1, no.1, p. 2.

51 Quoted in Robert J. Einhorn, April 26, 2006, "The US-India Nuclear Deal," Testimony of Robert J. Einhorn before the Senate Foreign Relations Committee, p. 5.

52 Quoted in Dr. Rashid Ahmad Khan, Winter 2007, "India-US Nuclear Deal: Implications For The Region," *IPRI Journal*, vol. VII, no. 1, p. 25.

53 Ramjet Devraj, March 6, 2006, "India Deal Makes USA Nuclear Proliferator." Available at www.commondreams.org/headlines06/0304-02htm, (accessed on 07 January 2014).

54 Quoted in Dr. Rashid Ahmad Khan, "India-US Nuclear Deal: Implications for The Region," *op. cit.*, p. 17.

55 K J M Varma, March 18, 2006, "India's Special Status Will Have Serious Implications for South Asia: Pakistan." Available at http://www.in.rediff.com/news/2006/mar/17ndeal4.htm, (accessed on 19 February 2014).

56 "Concerns Expressed at Indo-US Deal: Deterrence Satisfactory: NCA," 2006, *Dawn*, 13 April.

57 See the text of "Agreement for Cooperation Between the Government of India and the Government of United States of America Concerning Peaceful Uses of Nuclear Energy," pp. 6-7.

58 Adil Sultan Muhammad, July 2006, "Indo-US Civilian Nuclear Cooperation Agreement: Implications on South Asian Security Environment," The Henry L. Stimson Center, p. 2.

59 *Ibid.*, p. 13.

60 Robert J. Einhorn, "The US-India Nuclear Deal," *op. cit.*, p. 6.

61 Quoted in Sharif Shuja, 2007, "Pakistan Feels Jilted By US-India Nuclear Deal", *News Weekly*, 17 February.

62 "US-India Nuclear Deal." Available at http://www.reachingcriticalwill.org/resources/fact-sheets/critical-issues/5451-us-india-nuclear-deal, (accessed on 12 December 2013).

63 Quoted in Adil Muhammad Sultan, "Indo-US Civilian Nuclear Cooperation Agreement: Implications on South Asian Security Environment," op. cit., p. 19.

64 Carl Paddock, 2009, *India-US Nuclear Deal: Prospects and Implications*, New Delhi: Epitome Books, pp. 138-139.

65 Cited in K. Alan Kronstadt, September 8, 2005, "US-India

Bilateral Agreements in 2005," *CRS Report for Congress*, Congressional Research Service.

66 Kamal Mitra Chenoy and Anuradha M. Chenoy, September 1-7, 2007, "India's Foreign Policy Shifts and The Calculus of Power", *Economic and Political weekly*, vol. 42, no. 35, p. 3352.

67 Ashley J. Telis, 2005, "India As A New Global Power: An Action Agenda for The United States," Carnegie Endowment For International Peace, p. 35.

68 Nasrullah Mirza and M. Sadiq, January 2008, "Indo-US 123 Agreement: Impacts on Deterrence Stability in South Asia," Research Report 7, South Asian Strategic Stability Institute, p. 11

69 Quoted in Ananth Krishnan, 2013, "China-Pakistan Signed Deal on 1000 MW Reactor," *The Hindu*, 26 March.

70 For details see Information Circular 254/Rev.10/ Part 1.

71 For details see Daniel Horner, April 2011, "Chinese-Pakistani Reactor Deal Moves Ahead," *Arms Control Today*, Vol. 41 and Ashley J. Tellis, July 16, 2010, "The China-Pakistan Nuclear Deal: Separating Fact From Fiction", *Policy Outlook*, Carnegie Endowment for International Peace.

72 Mark Hibbs, April 27, 2010, "Pakistan Deal signals China's Growing Nuclear Assertiveness," *Nuclear Energy Brief*, Carnegie Endowment for International Peace.

73 Fred McGoldrick, January/February 2011, "The Road Ahead for Export Controls: Challenges for the Nuclear Suppliers Group," *Arms Control Today*, vol. 41.

74 Quoted in Howard Lafranchi, 2010, "US objects to China-Pakistan Nuclear Deal Hypocritical?," *The Christian Science Monitor*, 16 June.

75 Quoted in Howard Lafranchi, 2008, "Fallout of US-India Nuke Deal," *The Christian Science Monitor*, 24 October.

76 "Per Capita Energy Consumption." Available at http//: www.energyrealities.org/chapter/meeting-our-need/item/per-capita-energy-consumption/erp327B7c729A3B31D2B, (accessed on 22 January 2014).

77 "India's Energy Security: Key Issues impacting the India Oil and Gas sector," p. 3. Available at http://www.ey.com/Publication/vwLUAssets/Indias_energy_security/$FILE/India-s_energy_security.pdf, (accessed on 11 February 2014).

78 *Pratiyogta Darpan*, January 2014, pp. 1190-91.

79 Neeraj Jain, 2012, *Nuclear Energy: A Technology from Hell*, New Delhi: Aakar Books, pp. 238-239.

80 P. Garg, 2012, "Energy Scenario and Vision 2020 in India," *Journal of Sustainable Energy and Environment*, p.8

81 "Energy Statistics 2013", Report of Ministry of Statistics and Programme implementation, p. 11

82 "Securing Tomorrow's Energy Today: Policy and Regulations," February 2013, P. 11. Available at http://www.deloitte.com/assets/Dcom-India/Local%20Assets/Documents/IEC%202013/LongTerm Energy_Security.pdf, (accessed on 22 November 2013).

83 "Inventory of Geological Resources of Coal in India." Available at http://www.coal.nin.in/welcome.html, (accessed on 12 December 2013).

84 A.P.J. Abdul Kalam and Srijan Pal Singh, 2011, "Nuclear Power is our Gateway to a Prosperous Future," *The Hindu*, 6 November.

85 "Securing tomorrow's Energy Today: Policy and Regulations," *op. cit.*, P. 17.

86 "Indian Energy Sector: An Overview," (2012/13), *TERI Energy Data Directory and Yearbook*, p. 5.

87 "Securing tomorrow's Energy Today: Policy and Regulations," *op. cit.*, p.17.

88 "Energy Statistics 2013," *op. cit.*, P. 2.

89 "India's Energy Security: Key issues Impacting the Indian Oil and Gas Sector," *op. cit.*, p. 3

90 *Ibid.*

91 Report of U.S. Energy Information Administration. Available at http://www.eia.gov/countries/analysisbriefs/India/india.pdf, (accessed on 1 January 2014).

92 Asit Samadder and Surbhi Arora, "Natural Gas Scenario in India: An Overview," September 10, 2012, Available at http://ssrn.com/abstract=2144061, (accessed on 12 January 2014).

93 Praveen Saxena and Arun Kumar, October 2010, "Hydropower Development in India," p. 2. Available at http://www.org/Paper2010/TSA01.pdf, (accessed on 11 February 2014).

94 "Hydro Power Projects: Social and Environmental Perspective," pp. 19-20. Available at http://www.socialissuesindia.files.wordpress.com/2011/12/hydro-power-development-in-india.pdf, (accessed 21 January 2014).

95 M. Ramesh, 2013, "Hydro: Ambitions Watered Down," *The Hindu Business Line,* 1 September.

96 "Mapping India's Renewable Energy Growth potential: Status and Outlook 2013," p. 3. Available at www.ey.com/Publication/vwLUAssets/Mapping_Indias_Renewable_ Energy_growth_potential/$FILE/EY-Mapping-India's-Renewable-Energy-growth-potential.pdf, (accessed on 2 February 2014).

97 "India's Renewable Energy Sector: Potential and Investment opportunities," p. 4. Available at www.direc2010.gov.in/pdf/India's-RE-Sector-Potential-and.Investment-opportunities-SSM.pdf, (accessed on 23 December 2013).

98 Neeraj Jain, *Nuclear Energy: A Technology from Hell, op. cit.*, p. 258.

99 "Energy Statistics 2013," *op. cit.* p. 3.

100 "Mapping India's Renewable Energy Growth Potential: Status and Outlook 2013," *op. cit.*, p. 24.

101 Neeraj Jain, *Nuclear Energy: A Technology from Hell, op. cit.*, p. 266.

102 Sujay Mehdudia, 2013, "India poised to Become Global Solar Power: World Bank," *The Hindu*, 12 December.

103 *Ibid.*

104 "Solar Energy Ambitions Take Shape as Costs Tumble," January 15, 2014, *Times Of India. Available at:* http://www.timesonindia.indiatimes.com/business/india-business/Solar-energy-ambitions-take-shape-as-cost-tumble/articlesshow/28846123.cms, accessed on 25/1/2014.

105 Report of U.S. Energy Information Administration, *op. cit.*

106 "Mapping India's Renewable Energy Growth Potential," *op. cit.*, p. 21

107 Quoted in Neeraj Jain, *Nuclear Energy: A Technology from Hell, op. cit.*, p.3.

108 Sonika Gupta et al, 2007, "Assessing the Indo-US Deal on Civil Nuclear Cooperation: Managing Risks and Opportunities," *NIAS Report*, p. 1.

109 Zia Mian and M.V. Ramana, "Wrong Ends, Means and Needs: Behind the US Nuclear Deal with India," op. cit.

110 M.V. Ramana, March 2007, "Nuclear Power in India: Failed Past Dubious Future" in Henry Sokolski (ed.), *Gauging U.S.-India Strategic Cooperation*, Report of The Strategic Studies Institute, p. 80.

111 Praful Bidwai, July 30 - August 5, 2005, "A Deplorable Nuclear Bargain," *Economic and Political Weekly*, vol. 40, no. 31, p. 3364.

112 M.V Ramana, 2012, *The Power of Promise: Examining Nuclear Energy in India, New Delhi: Penguin*, p. 271.

113 "Indo-US Civil Nuclear Deal." Available at http://www.ukessays.com/essays/sciences/indo-us-civilian-nuclear deal.php, (accessed on 10 November 2013).

114 Neeraj Jain, *Nuclear Energy: A Technology from Hell, op. cit.*, p. 57.

115 *Ibid.* p. 67.

116 Vladimir Radyuhin, 2011, "Twenty five Years after Chernobyl," *The Hindu*, 26 April.

117 Suvrat Raju and M.V. Ramana, 2011, " Why Kundankulam is Untenable," *The Hindu*, 12 November.

118 Quoted in M.V. Ramana, *The Power of Promise: Examining Nuclear Energy in India, op.cit.*, p. 157.

119 Neeraj Jain, *Nuclear Energy: A Technology from Hel, op. cit.*, p. 99.

120 M. V. Ramana, March 2007, "Nuclear Power in India: Failed Past Dubious Future" in Henry Sokolski (ed.), *op. cit.*, P. 86.

121 Quoted in Praful Bidwai, "A Deplorable Nuclear Bargain," *op. cit.*, p. 3363.

122 S. Nagesh, 2013, "In Nuclear Slow Down Small is Better," *The Hindu*, 7 September.

123 Quoted in Neeraj Jain, *Nuclear Energy: A Technology from Hell, op. cit.*, p. 71.

124 S. Nagesh, "In Nuclear Slow Down Small is Better," *op. cit.*

125 Neeraj Jain, *Nuclear Energy: A Technology from Hell, op. cit.*, p.76.

5

Conclusion

India and the United States have experienced an era of frosty relationship for a long stretch of time. The main factor responsible for this cold relationship was the "nuclear issue" which not only impaired their relations in the nuclear field but had an adverse affect on every area of cooperation. The study reveals that the relations between the two sides after a smooth beginning in 1950s started straining in mid 1960s because of India's decision not to join Nuclear Non-Proliferation Treaty and deteriorated in 1974 when India detonated a nuclear device at Pokhran making US to terminate its relations with India. Thereafter, two sides remained at logger heads for decades over the nuclear issue. The process of rapprochement initiated after India's nuclear test to transform the relations between two sides could not realise the full potential of India-US bilateral cooperation because of its limited scope. No real attempt was made to settle down the "nuclear issue" which for decades had dogged the relations between two sides. However, with the end of Cold War and 9/11 crises, new realities came to the forefront which prompted both the countries to redefine their relationship. Number of initiatives were taken in this direction but the most important and historical initiative was the India-US Civil Nuclear Cooperation Initiative which has marked the beginning of a new era in India-US bilateral relationship. The Agreement has a strategic importance and is mutually beneficial to both the countries in a number of areas.

For India, the Agreement ends India's decades old nuclear isolation and marks an end to the technology denial regimes. The Agreement recognizes India as a de-facto nuclear weapon state and puts an end to her status as a pariah in global nuclear order. The Agreement has lifted global moratorium on nuclear trade and commerce with India and provides India access to nuclear material, equipment and dual use technology. Right after the signing of the Agreement India has singed civil nuclear cooperation agreements with a number of countries including Russia, France, Canada etc. Besides providing access to nuclear material, equipment and technology, the Agreement carries a promise of India's participation in international research projects like ITER (International Thermonuclear Experimental Reactor) and Generation IV International Forum. India's participation in these projects is likely to give a boost to her indigenous research and development in the field of nuclear energy.

The greatest strategic achievement for India is that the Agreement per se does not place any restraints on India's nuclear weapons programme; rather it can have a reverse effect. India can exploit the Agreement to augment the growth of its weapons programme. A report of the Congressional Research Service on the Agreement notes that: "There are no measures in this global partnership to restrain India's nuclear weapons programme."[1] The Agreement obligates India to place only civilian part of her nuclear programme under safeguards. The military part will continue to serve to her strategic weapons programme. As mentioned, the Agreement has lifted global moratorium on the supply of nuclear fuel to India. India can now freely import fuel from international market to feed its civilian nuclear power programme while indigenous reserves of uranium can be diverted towards the production of nuclear weapons. A former Indian intelligence official has said that the fuel supply from US and other suppliers "would free India's existing capacity to produce highly enriched uranium and plutonium for its nuclear weapons programme... Under the deal, India... have the capacity to produce 50 warheads a year."[2]

Under the Agreement, India has committed to continue its unilateral moratorium on nuclear testing and to negotiate a multilateral Fissile Material Cut off Treaty. These two commitments as such do not pose a big challenge for India's strategic nuclear programme. India has been observing unilateral moratorium on nuclear testing since 1998 and it can review this policy in the interest of her supreme national interests and the Agreement does not take away such a right from India.

The negotiations on FMCT have frozen/stuck for a number of years and no progress is being made on the issue because of serious differences between the members of Conference on Disarmament and it seems highly unlikely that the treaty will come to fruition in near future, giving ample time to India to accumulate enough fissile material to satisfy the requirements of minimum deterrence. The Agreement itself does not place any restraints on India's fissile material production. Instead, it calls up on both the parties not to interfere in each other's strategic nuclear programmes. So the argument of the opponents that the deal compromises India's strategic nuclear programme is not valid. It is true that the Hyde Act calls up on US to terminate nuclear cooperation with India if India detonates a nuclear device but the Agreement provides enough manoeuvring room to India to defend its nuclear test. Before the termination of the nuclear cooperation, Agreement calls up on the parties to hold consultations and take into account relevant factors like changed security environment or a similar action by other states which prompted an Indian test (Art. 14). So the terms of the termination are open to consultation and the party seeking the termination of the Agreement has to give due recognition to other party's security concerns before invoking such a right. As the study asserts that one of the instrumental factors behind the US offer of civil nuclear cooperation to India is counterbalancing China, the US policy makers cannot ignore the fact that further augmentation of India's nuclear arsenal will not only help India to develop a credible deterrent against China but will also shift strategic balance of the region in India's favour, thus serving a long term objective of US foreign policy. A Pakistani analyst has rightly observed that "if counterbalancing China is one minor objective of the US, it can hardly seek to further that end if it does not permit India to develop a credible nuclear deterrence towards China."[3] During the negotiations on the Agreement, US based think tanks and strategic analysts also stressed that no constraints must be placed on India's nuclear weapons programme to achieve broad US strategic interests in Asia. A noted strategic analyst Ashley J. Tellis noted that:

> *"By integrating India into the non-proliferation order at the cost of capping the size of its eventual nuclear deterrent [the US would] threaten to place New Delhi at a severe disadvantage vis-à-vis Beijing, a situation that could not only undermine Indian security but*

also the US interests in Asia in the face of prospective rise of Chinese power over the long term."[4]

Moreover, an amendment to the Agreement requiring termination of the nuclear supplies to India if India conducted a nuclear test was rejected by the Senate. All this indicates that US response to an Indian nuclear test will be greatly determined by geo-strategic considerations. The apprehensions that the deal would lead to the erosion of India's indigenous research and development in the field of nuclear energy are also invalid. The "separation plan" makes it explicitly clear that India would open only one part of its nuclear programme for IAEA inspectors and rights of IAEA would extend to only civilian part of India's nuclear programme. Agency has no right or mandate to see what is going on in other part of India's nuclear programme which has been kept outside the purview of IAEA to serve India's security interests.

The India-US Civil Nuclear Cooperation Agreement is seen as watershed event in the India-US bilateral relations because of its broad strategic outlook. The deal not only focuses on the cooperation between two sides in the field of nuclear energy but also encompasses cooperation across a broad spectrum including environment, climate change, terrorism, close military and strategic ties, intelligence sharing etc. But in spite of its great strategic importance, the deal is not a win-win situation for India. Notwithstanding the fact that for India Agreement has been instrumental in ending her decades long nuclear isolation and to win trust of the United States but India has emerged as loser on other fronts because of the shifting of goalposts by US and diluting the basic parameters of July 18, 2005 understanding.

In the first place, India has not been able to secure life time fuel supply assurances for the safeguarded reactors to avoid Tarapur like experience where US terminated fuel supplies to India because of her nuclear test. No doubt Agreement notes a very important commitment that US will "create necessary conditions for India to have assured and full access to fuel for its reactors" and will "support an Indian effort to develop a 'strategic reserve' of nuclear fuel to guard against any disruption of fuel supply over the life time of India's reactors." However, what has been recorded in the Agreement in the name of fuel supply assurances are only political commitments not legally binding on US. This was confirmed by US President George W. Bush in a letter written to US Congress. In the letter Bush had stated that:

> *"[T]he agreement records certain political commitments... The text of the agreement does not transform these political commitments into legally binding commitments because the agreement, like other U.S. agreements of this type is intended as a framework agreement."*[5]

US Congress while framing the Hyde Act also out rightly rejected these assurances. The Act does not permit US to help India to build up a "stockpile" of nuclear fuel to guard against any disruption of fuel supplies but instructs it to provide only that amount of fuel to India sufficient for "reasonable reactor operating requirements" (section 103(b)(10). If a situation arises in future where fuel supplies to India are disrupted either due to termination of the Agreement or because of an Indian action that is inconsistent with July 18, 2005 joint statement or the provisions of the Hyde Act, US will not terminate its own supplies but will also encourage other suppliers to stop fuel supplies to India (section 103(a)(6). Officials of US State Department have reiterated that under the Agreement US is obliged to restore fuel supplies to India if a disruption occurs due to market failures or any other similar reason. This was testified by Under Secretary Nicholas Burns before a US Congressional Committee:

> *"...we are determined to help India to try to ensure a reasonable steady supply of fuel and should disruptions arise, for example, [owing to] trade disputes, [when] a commercial firm fails to meet its requirements, then we are firmly determined to... meet those requirements to the fullest extent consistent with U.S. law."* [6]

Given the fact that India has very limited reserves of uranium not sufficient to keep its fleet of nuclear reactors operational for a long time, any disruption of fuel supplies would cripple India's nuclear power programme and would left nuclear installation worth billions of dollars idle. To avoid such a scenario India must ensure that any civil nuclear cooperation agreement signed with any country carries the guarantees of uninterrupted fuel supplies over the life time of civilian reactors.

Secondly, the promise of "full civil nuclear cooperation" as underlined in July 18, 2005 joint communiqué has been diluted to greater extent by excluding the transfer of enrichment and reprocessing technology to India. The Agreement makes the transfer of these technologies to India conditional up on an amendment to the Agreement. The officials of US State Department have made it clear that US has no intention to carry out an amendment to the Agreement to facilitate the transfer of such technologies to India because US does not consider such technologies necessary for the generation of nuclear power. During a debate on the Hyde Act, Senator Joseph Biden asserted that "these technologies [enrichment and reprocessing technologies] are all used to produce fissile materials... it is US policy not to sell them [Indians] these technologies, so this is a matter more of pride than substance, which I hope they deal with."[7] It is one of the basic principles of US non-proliferation policy not to transfer such technologies under 123 Agreements. The civil nuclear cooperation agreements signed between US and other non-nuclear weapon states also do not involve the transfer of such technologies. So, it will be an exaggeration to assume that India will be an exception. Moreover, India is also prohibited to seek the transfer of such technologies from other suppliers because of the latest changes made in NSG guidelines. The new NSG guidelines make it clear that the states parties to the NPT and having Comprehensive Safeguards Agreements in place can access such technologies and India does not fulfil any of these conditions. The new guidelines have subverted the letter and spirit of the waiver granted to India in 2008. To rectify the damage caused to the commitment of "full civil nuclear cooperation" India must adopt a tough diplomatic stand while negotiating civil nuclear agreements with NSG countries and must give them a clear message that their entry in to Indian market will depend on their cooperation with India in all aspects of nuclear fuel cycle including enrichment and reprocessing technologies.

Thirdly, the Agreement does not permit India to unilaterally remove its facilities from safeguards on security reasons. This is a serious departure from July 18, 2005 understanding which describes India as a "responsible state with advanced nuclear technology" and states that it should "acquire the same benefits and advantages as other such states." All the five nuclear weapon states have signed Voluntary Offer Safeguards

Agreements with IAEA which give them the right to withdraw facilities from safeguards on national security grounds. Contrary to this, the Safeguards Agreement signed between India and IAEA does not permit India to unilaterally withdraw the safeguarded material and facilities from safeguards. Safeguards on nuclear material and facilities can be terminated only when jointly determined by India and the Agency and when material and the facilities subject to safeguards are no longer relevant for safeguards. However, Indian officials claim that India has retained the right to withdraw her facilities from safeguards if fuel supplies are disrupted because one of the important bases for India to agree to permanent safeguards was the uninterrupted fuel supply, if such a condition ceases to exist India has a right to withdraw reactors from safeguards. This was asserted by former Chairman of India's Energy Commission Dr. Anil Kakodkar: "The permanence of safeguards is coupled with the perpetuity of supply of fuel. So, in case there is a disruption in the supply of fuel, the safeguards agreement will cease to hold."[8] But such claims have been rejected by the officials of IAEA who time and again have stressed on the permanent nature of the Safeguards Agreement. These conflicting claims can become a cause of confrontation between the two sides and the ambiguous language of the Safeguards Agreement can further exacerbate the situation. However, if a situation arises in future where India is compelled to withdraw her facilities from safeguards which can only happen if fuel supplies to India were terminated, there will be little manoeuvring room for India to defend her decision because IAEA has made its position very clear that the Agency will not compromise with the "perpetuity principle" of safeguards arrangement.

Fourth, what India is going to receive in the name

of nuclear fuel and technology is hedged with conditions because the waiver granted to India by NSG in September 2008 makes the transfer of nuclear fuel, equipment and technology to India conditional up on India's non-proliferation commitments recorded by then India's External Affairs Minister in his September 5 statement and the text of the waiver makes an explicit referenced to these commitments in paragraph 3 of the waiver. The waiver calls for a periodic review of these commitments and any non-compliance from Indian side can prompt NSG to hold consultations and act in accordance with paragraph 16 of the guidelines which calls for taking appropriate action including termination of nuclear supplies if there is violation of supplier/recipient understanding.[9]

Lastly, Agreement could have far reaching consequences for India's independent foreign policy. International relations work on the principle of quid pro quo and nothing comes free of cost particularly when asymmetry exists in terms of power equations between the partners engaging in a strategic partnership as Praful Bidwai, a noted journalist and writer has rightly observed:

> *"Partnership is no free lunch. The dominant power will lay down the terms of 'partnership' and 'responsibility'; the subordinate power must contortedly strive to adjust to the terms, sometimes at the enormous harm to its interests in some field or other."*[10]

This has come true to a large extent in the case of India-US strategic entente. Since the announcement of the nuclear deal in 2005, number of steps have been taken by India indicating that the country is getting aligned with US at the cost of her long cherished foreign policy goals. In September 2005 and in February 2006, India voted twice in favour of US sponsored motion at IAEA to refer Iran to United Nation's Security Council because of her controversial nuclear programme. India's vote against Iran at IAEA was in sharp contradiction to her long held position that Iran being a member of NPT has every right to all aspects of nuclear fuel cycle including enrichment and reprocessing. This change in Indian position made Senator Richard Lugar to say:

> *"We have already seen strategic benefits from our improving relationship with India. India's vote at IAEA on Iranian issue...demonstrate that New Delhi is able to adjust its traditional foreign policies and play a constructive role on international issues."*[11]

In supporting US on Iran's nuclear issue, India has fulfilled the requirements of section 103(b) (4) of the Hyde Act which calls up on US to secure India's full and active support in isolating Iran.

Besides Iran's nuclear issue, India also succumbed to US pressure on much ambitious Iran-Pakistan-India (IPI) gas pipeline project. The project would have gone a long way in quenching India's growing appetite for energy. Realising that the project could undermine US efforts to isolate Iran, Bush administration exerted pressure on India to quit the project. Besides being made to support US on its "Iran isolation efforts", India is also being drawn into highly controversial counter proliferation measures like Proliferation Security Initiative (PSI) and Container Security Initiative (CSI). The Hyde Act report submitted to US Congress pursuant to section 104(C) of the Act reveals that discussions between India and US on PSI are going on and in 2007, India attended PSI's three regional interdiction services as an official observer. Moreover, the Nuclear Proliferation Assessment Statement (NPAS) submitted by US President to Congress pursuant to section 123(a) of US Atomic Energy Act of 1954 as amended reveals that India has also agreed to Join CSI and is participating in other initiatives similar to CSI.[12] It must be noted that these are the issues which Indian Prime Minster once termed as "extraneous" and outside the parameters of July 18, 2005 understanding.

In an attempt to analyze the strategic implications of the Agreement for India, the study has raised several important issues. Among these two important and crucial issues are: the role which nuclear energy can play in meeting India's long term energy requirements, and implications of the Agreement for the stability of South Asian region. The proponents of the Civil Nuclear Cooperation Agreement have hailed the Agreement as a panacea for India's growing energy demands. The study, however, asserts that the nuclear energy can play a marginal role in meeting India's long term energy requirements. "Failed past" of India's nuclear sector and challenges associated with the development of nuclear energy make it an unattractive source of energy. It is also an expensive source of energy and huge investments made in this sector will make hostage the development of other sources of energy. It has been estimated that to produce 1MW of electricity through imported reactors would cost India around Rs 11 crore.[13] The DAE has set a target to generate 20,000 MW by 2020. To realise this target, India has to import about 10,000 MW off nuclear power because India's domestic reserves of uranium are sufficient to support 10,000 MW only. It would cost India about Rs 1,10,000 crore. The same figure if produced through thermal power plants or through hydropower plants would cost Rs 40,000 crore and 30,000 crore respectively.[14] So, it makes no financial sense to produce power through expensive means when other cheaper means are available.

Besides being expensive, it is also an unsafe source

of energy. Past record of global nuclear industry reveals that accidents with catastrophic consequences are inevitable with this complex technology. So why to spend on a dangerous and an expensive technology when there are other safe and cost effective ways to meet country's long term energy needs. India is blessed with a vast potential of renewable energy and efforts must be taken to expand this source. Increased energy efficiency measures, advanced combustion methods, restructuring country's coal industry, reduction in system loses could help to save many times more energy than currently produced through nuclear means. In India transmission and distribution loses very high and estimates put it over 32 per cent of the total generating capacity compared to global average of 10 percent.[15] According to studies conducted by some independent agencies these loses range between 40-50 per cent. According to the estimates of Planning Commission of India the system loses in India exceed 40 percent. A 5 per cent reduction in such loses could help to save 11-15 gigawatts of electricity by 2016 and an additional 5 percent reduction between 2016 and 2032 could help to save 25-39 gigawatts. So, with increased energy efficiency measures and reduction in system loses country could save 49-58 gigawatts by 2016 and about 67-106 gigawatts by 2032.[16]

So far as the implications of the Agreement for the South Asian region are concerned, it has been argued that the deal poses a great challenge to the stability of the region. The Agreement does not place any safeguards on India's nuclear weapons programme and has the potential to boost India's nuclear weapons capability. This can force Pakistan to take counter measures in view of the fact that the strategic nuclear programmes of the two countries are primarily driven by the security dilemma prevailing between the two countries. Pakistan has already made its position obvious that it will take every possible measure to maintain the strategic balance of the region.[17] Some studies have revealed that in its attempts to catch up with India, Pakistan since the announcement of Agreement is greatly diversifying its nuclear programme and is improving its nuclear forces both qualitatively and quantitatively.[18] It is a potentially dangerous situation and can trigger an arms race in the region. To allay the concerns of Pakistan, it would have been more logical for Bush administration to clinch a similar type of agreement with Pakistan with more stringent non-proliferation conditions. Such an approach would not have only served the interests of global non-proliferation regime but would also have served the objective of strategic stability in the region.[19] A strategically stable South Asia is not only in the interests of the stakeholders of the region but also in the larger interests of global peace and security.

REFERENCES

1 Cited in Jayshree Bajoria and Esther Pan, 5 November 2010, "The U.S.-India Nuclear Deal." Available at:www.cfr.org/india/us-india-nuclear-deal//p9663, (accessed on 18 March 2014).

2 Quoted in Adil Sultan Mohammad, July 2006, "India-US Civilian Nuclear Cooperation Agreement: Implications for South Asian Security Environment," The Henry L. Stimson Center, p. 13.

3 Shaukat Qadir, "An Outsider's View." Available at www.india-seminar.com/2006/560/560/560%20shaokat%20qadir.htm, (accessed on 18 March 2014)

4 Quoted in Mohammad Tyab Khan, "The Geopolitical Dimensions of Indo-US Nuclear Deal," *Policy Perspectives,* vol. 3, no. 2.

5 Cited in R. Ramachandran, October 25 - November 7, 2008, "Waiting Traps," *Frontline,* vol. 25, issue 22.

6 Quoted in *ibid*.

7 Quoted in Arun Shourie, 2006, "Facts Versus the Government's Fiction," *The Indian Express,* 22 December.

8 Quoted in "The IAEA Safeguards Agreement Explained." Available at www.rediff.com/news/2008/jul/15sd1.htm, (accessed on 19 March 2014)

9 Arun Shourie, 2008, "An Empty Claim," *The Indian Express*, 8 September.

10 Praful Bidwai, September 24 - October 7, 2005, "India's Nuclear Albatross," *Frontline*, vol. 25, issue 20.

11 Quoted in Prakash Karat, 2007, *Subordinate Allay: The Nuclear Deal and India-US Strategic Relations*, New Delhi: Left Word Books.

12 R. Ramachandran, September 27 - October 10, 2008, "The Truth, yet Again," *Frontline*, vol. 25, issue 20.

13 Jasjit Singh et al, October - December 2007, "Indo-US Nuclear Deal: A Debate," *Vikalpa*, vol. 32, no. 4, p. 101.

14 As per the estimates the cost of electricity produced through thermal and hydropower plants is about Rs 4 crore and 3 crore per Megawatt respectively. For details see *Ibid*.

15 Minhaz Merchant, 2012, "Powerless in India," *The Times of India*, 12 July.

16 Jhon Stephenson and Peter Tyman, March 2007, "Will the Civil Nuclear Cooperation Initiative Light India," in Henry Sokoloski (ed.), *Gauging U.S.-Indian Strategic Cooperation*, A report published by Strategic Studies Institute, p. 33

17 For details see "Indo-US Nuclear Deal: Speakers urge NSG to Adopt "Non-discriminatory, Criteria Based Approach." Available at www.thenewstribune.com/2013/06/27indo-us-nuclear-deal-speakers-urge-nsg-to-adopt-non-discriminatory-criteria-based-approach, (accessed on 19 March 2014) and Dr. Rashid Ahmad Khan, Winter 2007, "India-US Nuclear Deal: Implications for the Region," *IPRI Journal*, vol.VII, no.1

18 Zina Varelas, April 15, 2011, "Analyzing the 2005-2008 U.S.-India Civil Nuclear Cooperation Deal as a Model for Dealing with Nuclear Rising Powers," A Thesis submitted to the Faculty of the Graduate School of Arts and Sciences of Georgetown University in Partial fulfillment of the requirements for the degree Master of Arts in Security studies, pp. 36-38

19 K J M Varma, March 18, 2006, "India's Special Status to Affect South Asia." Available at www.rediff.com/news/2006/news/2006/mar/17ndeal14.htm, (accessed on 19 March 2014)

Annexures

1. INDIA - USA JOINT STATEMENT OF 18 JULY, 2005

Prime Minister Manmohan Singh and President Bush today declare their resolve to transform the relationship between their countries and establish a global partnership. As leaders of nations committed to the values of human freedom, democracy and rule of law, the new relationship between India and the United States will promote stability, democracy, prosperity and peace throughout the world. It will enhance our ability to work together to provide global leadership in areas of mutual concern and interest.

Building on their common values and interests, the two leaders resolve:

- To create an international environment conducive to promotion of democratic values and to strengthen democratic practices in societies which wish to become more open and pluralistic.
- To combat terrorism relentlessly. They applaud the active and vigorous counter terrorism co-operation between the two countries and support more international efforts in this direction. Terrorism is a

global scourge and the one we will fight everywhere. The two leaders strongly affirm their commitment to the conclusion by September of a UN comprehensive convention against international terrorism.

The Prime Minister's visit coincides with the completion of the Next Steps in Strategic Partnership (NSSP) initiative launched in January 2004. The two leaders agree that this provides the basis for expanding bilateral activities and commerce in space, civil nuclear energy and dual-use technology.

Drawing on their mutual vision for the U.S.-India relationship and our joint objectives as strong long-standing democracies, the two leaders agree on the following:

FOR THE ECONOMY

- Revitalize the U.S.-India Economic Dialogue and launch a CEO Forum to harness private sector energy and ideas to deepen the bilateral economic relationship.
- Support and accelerate economic growth in both countries through greater trade, investment and technology collaboration.
- Promote modernization of India's infrastructure as a pre-requisite for the continued growth of the Indian economy. As India enhances its investment climate, opportunities for investment will increase.
- Launch a U.S.-India Knowledge Initiative on Agriculture focused on promoting teaching, research, service and commercial linkages.

FOR ENERGY AND THE ENVIRONMENT

- Strengthen energy security and promote the

development of stable and efficient energy markets in India with a view to ensuring adequate, affordable energy supplies and conscious of the need for sustainable development. These issues will be addressed through the U.S.-India Energy Dialogue.

- Agree on the need to promote the imperatives of development and safeguarding the environment, commit to developing and deploying cleaner, more efficient, affordable and diversified energy technologies.

FOR DEMOCRACY AND DEVELOPMENT

- Develop and support through the new U.S.-India Global Democracy Initiative in countries that seek such assistance, institutions and resources that strengthen the foundations that make democracies credible and effective. India and the U.S. will work together to strengthen democratic practices and capacities and contribute to the new U.N. Democracy Fund.
- Commit to strengthen co-operation and combat HIV/AIDs at a global level through an initiative that mobilizes private sector and government resources, knowledge and expertise.

FOR NON-PROLIFERATION AND SECURITY

- Express satisfaction at the New Framework for the U.S.-India Defense Relationship as a basis for future co-operation including in the field of defense technology.
- Commit to play a leading role in international efforts to prevent the proliferation of Weapons of Mass Destruction. The U.S. welcomed the adoption by India of legislation on WMD (Prevention of Unlawful Activities Bill).

- Launch a new U.S.-India Disaster Relief Initiative that builds on the experience of the Tsunami Core Group to strengthen co-operation to prepare for and conduct disaster relief operations.

FOR HIGH-TECHNOLOGY AND SPACE

- Sign a Science and Technology Framework Agreement building on the U.S.–India High-Technology Co-operation Group (HTCG) to provide for joint research and training and the establishment of public-private partnerships.
- Build closer ties in space exploration, satellite navigation and launch and in the commercial space arena through mechanisms such as the U.S.-India Working Group on Civil Space Co-operation.
- Building on the strengthened non-proliferation commitments undertaken in the NSSP to remove certain Indian organizations from the Department of Commerce's Entity List.

Recognizing the significance of civilian nuclear energy for meeting growing global energy demands in a cleaner and more efficient manner, the two leaders discussed India's plans to develop its civilian nuclear energy program.

President Bush conveyed his appreciation to the Prime Minister over India's strong commitment to preventing WMD proliferation and stated that as a responsible state with advanced nuclear technology, India should acquire the same benefits and advantages as other such states. The President told the Prime Minister that he will work to achieve full civil nuclear energy co-operation with India as it realizes its goals of promoting nuclear power and achieving energy security. The President would also seek agreement from Congress to adjust U.S. laws and policies and the United States will work with friends and allies to adjust international regimes to enable full civil nuclear energy co-operation and trade with India including but not limited to expeditious consideration of fuel supplies for safeguarded nuclear reactors at Tarapur. In the meantime, the United States will encourage its partners to also consider this request expeditiously. India has expressed its interest in ITER and a willingness to contribute. The United States will consult with its partners considering India's participation. The United States will consult with the other participants in the Generation IV International Forum with a view toward India's inclusion.

The Prime Minister conveyed that for his part, India would reciprocally agree that it would be ready to assume the same responsibilities and practices and acquire the same benefits and advantages as other leading countries with advanced nuclear technology such as the United States. These responsibilities and practices consist of identifying and separating civilian and military nuclear facilities and programs in a phased manner and filing a declaration regarding its civilians facilities with the International Atomic Energy Agency (IAEA), taking a decision to place voluntarily its civilian nuclear facilities under IAEA safeguards, signing and adhering to an Additional Protocol with respect to civilian nuclear facilities, continuing India's unilateral moratorium on nuclear testing, working with the United States for the conclusion of a multilateral Fissile Material Cut Off Treaty, refraining from transfer of enrichment and reprocessing technologies to states that do not have them and supporting international efforts to limit their spread; and ensuring that the necessary steps have been taken to secure nuclear materials and technology through comprehensive export control legislation and through harmonization and adherence to Missile Technology

Control Regime (MTCR) and Nuclear Suppliers Group (NSG) guidelines.

The President welcomed the Prime Minister's assurance. The two leaders agreed to establish a working group to undertake on a phased basis in the months ahead the necessary actions mentioned above to fulfil these commitments. The President and Prime Minister also agreed that they would review this progress when the President visits India in 2006.

The two leaders also reiterated their commitment that their countries would play a leading role in international efforts to prevent the proliferation of weapons of mass destruction, including nuclear, chemical, biological and radiological weapons.

In light of this closer relationship and the recognition of India's growing role in enhancing regional and global security, the Prime Minister and the President agree that international institutions must fully reflect changes in the global scenario that have taken place since 1945. The President reiterated his view that international institutions are going to have to adapt to reflect India's central and growing role. The two leaders state their expectations that India and the United States will strengthen their co-operation in global forums.

Prime Minister Manmohan Singh thanks President Bush for the warmth of his reception and the generosity of his hospitality. He extends an invitation to President Bush to visit India at his convenience and the President accepts that invitation.

2. AGREEMENT FOR COOPERATION BETWEEN THE GOVERNMENT OF THE UNITED STATES OF AMERICA AND THE GOVERNMENT OF INDIA CONCERNING PEACEFUL USES OF NUCLEAR ENERGY (123 AGREEMENT)

The Government of India and the Government of the United States of America, hereinafter referred to as the Parties, RECOGNIZING the significance of civilian nuclear energy for meeting growing global energy demands in a cleaner and more efficient manner;

DESIRING to cooperate extensively in the full development and use of nuclear energy for peaceful purposes as a means of achieving energy security, on a stable, reliable and predictable basis;

WISHING to develop such cooperation on the basis of mutual respect for sovereignty, non-interference in each other's internal affairs, equality, mutual benefit, reciprocity and with due respect for each other's nuclear programmes;

DESIRING to establish the necessary legal framework and basis for cooperation concerning peaceful uses of nuclear energy;

AFFIRMING that cooperation under this Agreement is between two States possessing advanced nuclear technology, both Parties having the same benefits and advantages, both committed to preventing WMD proliferation;

NOTING the understandings expressed in the India - U.S. Joint Statement of July 18, 2005 to enable full civil nuclear energy cooperation with India covering aspects of the associated nuclear fuel cycle;

AFFIRMING their support for the objectives of the International Atomic Energy Agency (IAEA) and its safeguards system, as applicable to India and the United States of America, and its importance in ensuring that international cooperation in development and use of nuclear energy for peaceful purposes is carried out under arrangements that will not contribute to the proliferation of nuclear weapons or other nuclear explosive devices;

NOTING their respective commitments to safety and security of peaceful uses of nuclear energy, to adequate physical protection of nuclear material and effective national export controls;

MINDFUL that peaceful nuclear activities must be undertaken with a view to protecting the environment;

MINDFUL of their shared commitment to preventing the proliferation of weapons of mass destruction; and

DESIROUS of strengthening the strategic partnership between them;

Have agreed on the following:

ARTICLE 1 – DEFINITIONS

For the purposes of this Agreement:

(A) "By-product material" means any radioactive material (except special fissionable material) yielded in or made radioactive by exposure to the radiation incident to the process of producing or utilizing special fissionable material. By-product material shall not be subject to safeguards or any other form of verification under this Agreement, unless it has been decided otherwise by prior mutual agreement in writing between the two Parties.

(B) "Component" means a component part of equipment, or other item so designated by agreement of the Parties.

(C) "Conversion" means any of the normal operations in the nuclear fuel cycle, preceding fuel fabrication and excluding enrichment, by which uranium is transformed from one chemical form to another - for example, from uranium hexafluoride (UF6) to uranium dioxide (UO2) or from uranium oxide to metal.

(D) "Decommissioning" means the actions taken at the end of a facility's useful life to retire the facility from service in the manner that provides adequate protection for the health and safety of the decommissioning workers and the general public, and for the environment. These actions can range from closing down the facility and a minimal removal of nuclear material coupled with continuing maintenance and surveillance, to a complete removal of residual radioactivity in excess of levels acceptable for unrestricted use of the facility and its site.

(E) "Dual-Use Item" means a nuclear related item which has a technical use in both nuclear and non-nuclear applications.

(F) "Equipment" means any equipment in nuclear operation including reactor, reactor pressure vessel, reactor fuel charging and discharging equipment, reactor control rods, reactor pressure tubes, reactor primary coolant pumps, zirconium tubing, equipment for fuel fabrication and any other item so designated by the Parties.

(G) "High enriched uranium" means uranium enriched to twenty percent or greater in the isotope 235.

(H) "Information" means any information that is not in the public domain and is transferred in any form pursuant to this Agreement and so designated and documented in hard copy or digital form by mutual agreement by the Parties that it shall be subject to this Agreement, but will cease to be information whenever the Party transferring the information or any third party legitimately releases it into the public domain.

(I) "Low enriched uranium" means uranium enriched to less than twenty percent in the isotope 235.

(J) "Major critical component" means any part or group of parts essential to the operation of a sensitive nuclear facility or heavy water production facility.

(K) "Non-nuclear material" means heavy water, or any other material suitable for use in a reactor to slow down high velocity neutrons and increase the likelihood of further fission, as may be jointly designated by the appropriate authorities of the Parties.

(L) "Nuclear material" means (1) source material and (2) special fissionable material. "Source material" means uranium containing the mixture of isotopes occurring in nature; uranium depleted in the isotope 235; thorium; any of the foregoing in the form of metal, alloy, chemical compound, or concentrate; any other material containing one or more of the foregoing in such concentration as the Board of Governors of the IAEA shall from time to time determine; and such other materials as the Board of Governors of the IAEA may determine or as may be agreed by the appropriate authorities of both Parties. "Special fissionable material" means plutonium, uranium-233, uranium enriched in the isotope 233 or 235, any substance containing one or more of the foregoing, and such other substances as the Board of Governors of the IAEA may determine or as may be agreed by the appropriate authorities of both Parties. "Special fissionable material" does not include "source material". Any determination by the Board of Governors of the IAEA under Article XX of that Agency's Statute or otherwise that amends the list of materials considered to be "source material" or "special fissionable material" shall only have effect under this Agreement when both Parties to this Agreement have informed each other in writing that they accept such amendment.

(M) "Peaceful purposes" include the use of information, nuclear material, equipment or components in such fields as research, power generation, medicine, agriculture and industry, but do not include use in, research on, or development of any nuclear explosive device or any other military purpose. Provision of power for a military base drawn from any power network, production of radioisotopes to be used for medical purposes in military environment for diagnostics, therapy and sterility assurance, and other similar purposes as may be mutually agreed by the Parties shall not be regarded as military purpose.

(N) "Person" means any individual or any entity subject to the territorial jurisdiction of either Party but does not include the Parties.

(O) "Reactor" means any apparatus, other than a nuclear weapon or other nuclear explosive device, in which a self-sustaining fission chain reaction is maintained by utilizing uranium, plutonium, or thorium or any combination thereof.

(P) "Sensitive nuclear facility" means any facility designed or used primarily for uranium enrichment, reprocessing of nuclear fuel, or fabrication of nuclear fuel containing plutonium.

(Q) "Sensitive nuclear technology" means any information that is not in the public domain and that is important to the design, construction, fabrication, operation, or maintenance of any sensitive nuclear facility, or other such information that may be so designated by agreement of the Parties.

ARTICLE 2 - SCOPE OF COOPERATION

1. The Parties shall cooperate in the use of nuclear energy for peaceful purposes in accordance with the provisions of this Agreement. Each Party shall implement this Agreement in accordance with its respective applicable treaties, national laws, regulations, and license requirements concerning the use of nuclear energy for peaceful purposes.
2. The purpose of the Agreement being to enable full civil nuclear energy cooperation between the Parties, the Parties may pursue cooperation in all relevant areas to include, but not limited to, the following:
   a. Advanced nuclear energy research and development in such areas as may be agreed between the Parties;
   b. Nuclear safety matters of mutual interest and competence, as set out in Article 3;
   c. Facilitation of exchange of scientists for visits, meetings, symposia and collaborative research;
   d. Full civil nuclear cooperation activities covering nuclear reactors and aspects of the associated nuclear fuel cycle including technology transfer on an industrial or commercial scale between the Parties or authorized persons;
   e. Development of a strategic reserve of nuclear fuel to guard against any disruption of supply

over the lifetime of India's reactors;

f. Advanced research and development in nuclear sciences including but not limited to biological research, medicine, agriculture and industry, environment and climate change;

g. Supply between the Parties, whether for use by or for the benefit of the Parties or third countries, of nuclear material;

h. Alteration in form or content of nuclear material as provided for in Article 6;

i. Supply between the Parties of equipment, whether for use by or for the benefit of the Parties or third countries;

j. Controlled thermonuclear fusion including in multilateral projects; and

k. Other areas of mutual interest as may be agreed by the Parties.

3. Transfer of nuclear material, non-nuclear material, equipment, components and information under this Agreement may be undertaken directly between the Parties or through authorized persons. Such transfers shall be subject to this Agreement and to such additional terms and conditions as may be agreed by the Parties. Nuclear material, non-nuclear material, equipment, components and information transferred from the territory of one Party to the territory of the other Party, whether directly or through a third country, will be regarded as having been transferred pursuant to this Agreement only upon confirmation, by the appropriate authority of the recipient Party to the appropriate authority of the supplier Party that such items both will be subject to the Agreement and have been received by the recipient Party.

4. The Parties affirm that the purpose of this Agreement is to provide for peaceful nuclear cooperation and not to affect the unsafeguarded nuclear activities of either Party. Accordingly, nothing in this Agreement shall be interpreted as affecting the rights of the Parties to use for their own purposes nuclear material, non-nuclear material, equipment, components, information or technology produced, acquired or developed by them independent of any nuclear material, non-nuclear material, equipment, components, information or technology transferred to them pursuant to this Agreement. This Agreement shall be implemented in a manner so as not to hinder or otherwise interfere with any other activities involving the use of nuclear material, non-nuclear material, equipment, components, information or technology and military nuclear facilities produced, acquired or developed by them independent of this Agreement for their own purposes.

ARTICLE 3 - TRANSFER OF INFORMATION

1. Information concerning the use of nuclear energy for peaceful purposes may be transferred between the Parties. Transfers of information may be accomplished through reports, data banks and computer programs and any other means mutually agreed to by the Parties. Fields that may be covered include, but shall not be limited to, the following:
   a. Research, development, design, construction, operation, maintenance and use of reactors, reactor experiments, and decommissioning;
   b. The use of nuclear material in physical, chemical, radiological and biological research, medicine, agriculture and industry;

c. Fuel cycle activities to meet future world-wide civil nuclear energy needs, including multilateral approaches to which they are parties for ensuring nuclear fuel supply and appropriate techniques for management of nuclear wastes;

d. Advanced research and development in nuclear science and technology;

e. Health, safety, and environmental considerations related to the foregoing;

f. Assessments of the role nuclear power may play in national energy plans;

g. Codes, regulations and standards for the nuclear industry;

h. Research on controlled thermonuclear fusion including bilateral activities and contributions toward multilateral projects such as the International Thermonuclear Experimental Reactor (ITER); and

i. Any other field mutually agreed to by the Parties.

2. Cooperation pursuant to this Article may include, but is not limited to, training, exchange of personnel, meetings, exchange of samples, materials and instruments for experimental purposes and a balanced participation in joint studies and projects.

3. This Agreement does not require the transfer of any information regarding matters outside the scope of this Agreement, or information that the Parties are not permitted under their respective treaties, national laws, or regulations to transfer.

4. Restricted Data, as defined by each Party, shall not be transferred under this Agreement.

ARTICLE 4 - NUCLEAR TRADE

1. The Parties shall facilitate nuclear trade between themselves in the mutual interests of their respective industry, utilities and consumers and also, where appropriate, trade between third countries and either Party of items obligated to the other Party. The Parties recognize that reliability of supplies is essential to ensure smooth and uninterrupted operation of nuclear facilities and that industry in both the Parties needs continuing reassurance that deliveries can be made on time in order to plan for the efficient operation of nuclear installations.
2. Authorizations, including export and import licenses as well as authorizations or consents to third parties, relating to trade, industrial operations or nuclear material movement should be consistent with the sound and efficient administration of this Agreement and should not be used to restrict trade. It is further agreed that if the relevant authority of the concerned Party considers that an application cannot be processed within a two month period it shall immediately, upon request, provide reasoned information to the submitting Party. In the event of a refusal to authorize an application or a delay exceeding four months from the date of the first application the Party of the submitting persons or undertakings may call for urgent consultations under Article 13 of this Agreement, which shall take place at the earliest opportunity and in any case not later than 30 days after such a request.

ARTICLE 5 - TRANSFER OF NUCLEAR MATERIAL, NON-NUCLEAR MATERIAL, EQUIPMENT, COMPONENTS AND RELATED TECHNOLOGY

1. Nuclear material, non-nuclear material, equipment

and components may be transferred for applications consistent with this Agreement. Any special fissionable material transferred under this Agreement shall be low enriched uranium, except as provided in paragraph 5.

2. Sensitive nuclear technology, heavy water production technology, sensitive nuclear facilities, heavy water production facilities and major critical components of such facilities may be transferred under this Agreement pursuant to an amendment to this Agreement. Transfers of dual-use items that could be used in enrichment, reprocessing or heavy water production facilities will be subject to the Parties' respective applicable laws, regulations and license policies.

3. Natural or low enriched uranium may be transferred for use as fuel in reactor experiments and in reactors, for conversion or fabrication, or for such other purposes as may be agreed to by the Parties.

4. The quantity of nuclear material transferred under this Agreement shall be consistent with any of the following purposes: use in reactor experiments or the loading of reactors, the efficient and continuous conduct of such reactor experiments or operation of reactors for their lifetime, use as samples, standards, detectors, and targets, and the accomplishment of other purposes as may be agreed by the Parties.

5. Small quantities of special fissionable material may be transferred for use as samples, standards, detectors, and targets, and for such other purposes as the Parties may agree.

6. (a) The United States has conveyed its commitment to the reliable supply of fuel to India. Consistent

with the July 18, 2005, Joint Statement, the United States has also reaffirmed its assurance to create the necessary conditions for India to have assured and full access to fuel for its reactors. As part of its implementation of the July 18, 2005, Joint Statement the United States is committed to seeking agreement from the U.S. Congress to amend its domestic laws and to work with friends and allies to adjust the practices of the Nuclear Suppliers Group to create the necessary conditions for India to obtain full access to the international fuel market, including reliable, uninterrupted and continual access to fuel supplies from firms in several nations.

(b) To further guard against any disruption of fuel supplies, the United States is prepared to take the following additional steps:

i) The United States is willing to incorporate assurances regarding fuel supply in the bilateral U.S.-India agreement on peaceful uses of nuclear energy under Section 123 of the U.S. Atomic Energy Act, which would be submitted to the U.S. Congress.

ii) The United States will join India in seeking to negotiate with the IAEA an India-specific fuel supply agreement.

iii) The United States will support an Indian effort to develop a strategic reserve of nuclear fuel to guard against any disruption of supply over the lifetime of India's reactors.

iv) If despite these arrangements, a disruption of fuel supplies to India occurs, the United States and India would jointly convene a group of friendly supplier countries to include countries such as Russia, France and the United Kingdom to pursue such measures as would restore fuel supply to India.

(c) In light of the above understandings with the United States, an India-specific safeguards agreement will be negotiated between India and the IAEA providing for safeguards to guard against withdrawal of safeguarded nuclear material from civilian use at any time as well as providing for corrective measures that India may take to ensure uninterrupted operation of its civilian nuclear reactors in the event of disruption of foreign fuel supplies. Taking this into account, India will place its civilian nuclear facilities under India-specific safeguards in perpetuity and negotiate an appropriate safeguards agreement to this end with the IAEA.

ARTICLE 6 - NUCLEAR FUEL CYCLE ACTIVITIES

In keeping with their commitment to full civil nuclear cooperation, both Parties, as they do with other states with advanced nuclear technology, may carry out the following nuclear fuel cycle activities:

i) Within the territorial jurisdiction of either Party, enrichment up to twenty percent in the isotope 235 of uranium transferred pursuant to this Agreement, as well as of uranium used in or produced through the use of equipment so transferred, may be carried out.

ii) Irradiation within the territorial jurisdiction of either Party of plutonium, uranium-233, high enriched uranium and irradiated nuclear material transferred pursuant to this Agreement or used in or produced through the use of non-nuclear material, nuclear material or equipment so transferred may be carried out.

iii) With a view to implementing full civil nuclear cooperation as envisioned in the Joint Statement of the Parties of July 18, 2005, the Parties grant each other consent to reprocess or otherwise alter in form or content nuclear material transferred pursuant to this Agreement and nuclear material and by-product material used in or produced through the use of nuclear material, non-nuclear material, or equipment so transferred. To bring these rights into effect, India will establish a new national reprocessing facility dedicated to reprocessing safeguarded nuclear material under IAEA safeguards and the Parties will agree on arrangements and procedures under which such reprocessing or other alteration in form or content will take place in this new facility. Consultations on arrangements and procedures will begin within six months of a request by either Party and will be concluded within one year. The Parties agree on the application of IAEA safeguards to all facilities concerned with the above activities. These arrangements and procedures shall include provisions with respect to physical protection standards set out in Article 8, storage standards set out in Article 7, and environmental protections set forth in Article 11 of this Agreement, and such other provisions as may be agreed by the Parties. Any special fissionable material that may be separated may only be utilized in national facilities under IAEA safeguards.

iv) Post-irradiation examination involving chemical dissolution or separation of irradiated nuclear material transferred pursuant to this Agreement or irradiated nuclear material used in or produced through the use of non-nuclear material, nuclear material or equipment so transferred may be carried out.

ARTICLE 7 - STORAGE AND RETRANSFERS

1. Plutonium and uranium 233 (except as either may be contained in irradiated fuel elements), and high enriched uranium, transferred pursuant to this Agreement or used in or produced through the use of material or equipment so transferred, may be stored in facilities that are at all times subject, as a minimum, to the levels of physical protection that are set out in IAEA document INFCIRC 225/REV 4 as it may be revised and accepted by the Parties. Each Party shall record such facilities on a list, made available to the other Party. A Party's list shall be held confidential if that Party so requests. Either Party may make changes to its list by notifying the other Party in writing and receiving a written acknowledgement. Such acknowledgement shall be given no later than thirty days after the receipt of the notification and shall be limited to a statement that the notification has been received. If there are grounds to believe that the provisions of this sub-Article are not being fully complied with, immediate consultations may be called for. Following upon such consultations, each Party shall ensure by means of such consultations that necessary remedial measures are taken immediately. Such measures shall be sufficient to restore the levels of physical protection referred to above at the facility in question. However, if the Party on whose territory the nuclear material in question is stored determines that such measures are not feasible, it will shift the nuclear material to another appropriate, listed facility it identifies.

2. Nuclear material, non-nuclear material, equipment, components, and information transferred pursuant to this Agreement and any special fissionable material produced through the use of nuclear material, non-nuclear material or equipment so transferred shall not be transferred or re-transferred to unauthorized persons or, unless the Parties agree, beyond the recipient Party's territorial jurisdiction.

ARTICLE 8 - PHYSICAL PROTECTION

1. Adequate physical protection shall be maintained with respect to nuclear material and equipment transferred pursuant to this Agreement and nuclear material used in or produced through the use of nuclear material, non-nuclear material or equipment so transferred.
2. To fulfil the requirement in paragraph 1, each Party shall apply measures in accordance with (i) levels of physical protection at least equivalent to the recommendations published in IAEA document INFCIRC/225/Rev.4 entitled "The Physical Protection of Nuclear Material and Nuclear Facilities," and in any subsequent revisions of that document agreed to by the Parties, and (ii) the provisions of the 1980 Convention on the Physical Protection of Nuclear Material and any amendments to the Convention that enter into force for both Parties.
3. The Parties will keep each other informed through diplomatic channels of those agencies or authorities having responsibility for ensuring that levels of physical protection for nuclear material in their territory or under their jurisdiction or control are adequately met and having responsibility for coordinating response and recovery operations in the event of unauthorized use or handling of

material subject to this Article. The Parties will also keep each other informed through diplomatic channels of the designated points of contact within their national authorities to cooperate on matters of out-of-country transportation and other matters of mutual concern.

4. The provisions of this Article shall be implemented in such a manner as to avoid undue interference in the Parties' peaceful nuclear activities and so as to be consistent with prudent management practices required for the safe and economic conduct of their peaceful nuclear programs.

ARTICLE 9 - PEACEFUL USE

Nuclear material, equipment and components transferred pursuant to this Agreement and nuclear material and by-product material used in or produced through the use of any nuclear material, equipment, and components so transferred shall not be used by the recipient Party for any nuclear explosive device, for research on or development of any nuclear explosive device or for any military purpose.

ARTICLE 10 - IAEA SAFEGUARDS

1. Safeguards will be maintained with respect to all nuclear materials and equipment transferred pursuant to this Agreement, and with respect to all special fissionable material used in or produced through the use of such nuclear materials and equipment, so long as the material or equipment remains under the jurisdiction or control of the cooperating Party.
2. Taking into account Article 5.6 of this Agreement, India agrees that nuclear material and equipment transferred to India by the United States of America

pursuant to this Agreement and any nuclear material used in or produced through the use of nuclear material, non-nuclear material, equipment or components so transferred shall be subject to safeguards in perpetuity in accordance with the India-specific Safeguards Agreement between India and the IAEA [***identifying data***] and an Additional Protocol, when in force.

3. Nuclear material and equipment transferred to the United States of America pursuant to this Agreement and any nuclear material used in or produced through the use of any nuclear material, non-nuclear material, equipment, or components so transferred shall be subject to the Agreement between the United States of America and the IAEA for the application of safeguards in the United States of America, done at Vienna November 18, 1977, which entered into force on December 9, 1980, and an Additional Protocol, when in force.

4. If the IAEA decides that the application of IAEA safeguards is no longer possible, the supplier and recipient should consult and agree on appropriate verification measures.

5. Each Party shall take such measures as are necessary to maintain and facilitate the application of IAEA safeguards in its respective territory provided for under this Article.

6. Each Party shall establish and maintain a system of accounting for and control of nuclear material transferred pursuant to this Agreement and nuclear material used in or produced through the use of any material, equipment, or components so transferred: The procedures applicable to India shall be those set forth in the India-specific Safeguards Agreement referred to in Paragraph 2 of this Article.

7. Upon the request of either Party, the other Party shall report or permit the IAEA to report to the requesting Party on the status of all inventories of material subject to this Agreement.

8. The provisions of this Article shall be implemented in such a manner as to avoid hampering, delay, or undue interference in the Parties' peaceful nuclear activities and so as to be consistent with prudent management practices required for the safe and economic conduct of their peaceful nuclear programs.

ARTICLE 11 - ENVIRONMENTAL PROTECTION

The Parties shall cooperate in following the best practices for minimizing the impact on the environment from any radioactive, chemical or thermal contamination arising from peaceful nuclear activities under this Agreement and in related matters of health and safety.

ARTICLE 12 - IMPLEMENTATION OF THE AGREEMENT

1. This Agreement shall be implemented in a manner designed:
   a) to avoid hampering or delaying the nuclear activities in the territory of either Party;
   b) to avoid interference in such activities;
   c) to be consistent with prudent management practices required for the safe conduct of such activities; and
   d) to take full account of the long term requirements of the nuclear energy programs of the Parties.

2. The provisions of this Agreement shall not be used to:
   a) secure unfair commercial or industrial

advantages or to restrict trade to the disadvantage of persons and undertakings of either Party or hamper their commercial or industrial interests, whether international or domestic;

b) interfere with the nuclear policy or programs for the promotion of the peaceful uses of nuclear energy including research and development; or

c) impede the free movement of nuclear material, non nuclear material and equipment supplied under this Agreement within the territory of the Parties.

3. When execution of an agreement or contract pursuant to this Agreement between Indian and United States organizations requires exchanges of experts, the Parties shall facilitate entry of the experts to their territories and their stay therein consistent with national laws, regulations and practices. When other cooperation pursuant to this Agreement requires visits of experts, the Parties shall facilitate entry of the experts to their territory and their stay therein consistent with national laws, regulations and practices.

ARTICLE 13 - CONSULTATIONS

1. The Parties undertake to consult at the request of either Party regarding the implementation of this Agreement and the development of further cooperation in the field of peaceful uses of nuclear energy on a stable, reliable and predictable basis. The Parties recognize that such consultations are between two States with advanced nuclear technology, which have agreed to assume the same responsibilities and practices and acquire the same

benefits and advantages as other leading countries with advanced nuclear technology.

2. Each Party shall endeavour to avoid taking any action that adversely affects cooperation envisaged under Article 2 of this Agreement. If either Party at any time following the entry into force of this Agreement does not comply with the provisions of this Agreement, the Parties shall promptly hold consultations with a view to resolving the matter in a way that protects the legitimate interests of both Parties, it being understood that rights of either Party under Article 16.2 remain unaffected.

3. Consultations under this Article may be carried out by a Joint Committee specifically established for this purpose. A Joint Technical Working Group reporting to the Joint Committee will be set up to ensure the fulfilment of the requirements of the Administrative Arrangements referred to in Article 17.

ARTICLE 14 - TERMINATION AND CESSATION OF COOPERATION

1. Either Party shall have the right to terminate this Agreement prior to its expiration on one year's written notice to the other Party. A Party giving notice of termination shall provide the reasons for seeking such termination. The Agreement shall terminate one year from the date of the written notice, unless the notice has been withdrawn by the providing Party in writing prior to the date of termination.

2. Before this Agreement is terminated pursuant to paragraph 1 of this Article, the Parties shall consider the relevant circumstances and promptly hold consultations, as provided in Article 13, to address the reasons cited by the Party seeking termination. The Party seeking termination has the right to cease

further cooperation under this Agreement if it determines that a mutually acceptable resolution of outstanding issues has not been possible or cannot be achieved through consultations. The Parties agree to consider carefully the circumstances that may lead to termination or cessation of cooperation. They further agree to take into account whether the circumstances that may lead to termination or cessation resulted from a Party's serious concern about a changed security environment or as a response to similar actions by other States which could impact national security.

3. If a Party seeking termination cites a violation of this Agreement as the reason for notice for seeking termination, the Parties shall consider whether the action was caused inadvertently or otherwise and whether the violation could be considered as material. No violation may be considered as being material unless corresponding to the definition of material violation or breach in the Vienna Convention on the Law of Treaties. If a Party seeking termination cites a violation of an IAEA safeguards agreement as the reason for notice for seeking termination, a crucial factor will be whether the IAEA Board of Governors has made a finding of non-compliance.

4. Following the cessation of cooperation under this Agreement, either Party shall have the right to require the return by the other Party of any nuclear material, equipment, non-nuclear material or components transferred under this Agreement and any special fissionable material produced through their use. A notice by a Party that is invoking the right of return shall be delivered to the other Party on or before the date of termination of this

Agreement. The notice shall contain a statement of the items subject to this Agreement as to which the Party is requesting return. Except as provided in provisions of Article 16.3, all other legal obligations pertaining to this Agreement shall cease to apply with respect to the nuclear items remaining on the territory of the Party concerned upon termination of this Agreement.

5. The two Parties recognize that exercising the right of return would have profound implications for their relations. If either Party seeks to exercise its right pursuant to paragraph 4 of this Article, it shall, prior to the removal from the territory or from the control of the other Party of any nuclear items mentioned in paragraph 4, undertake consultations with the other Party. Such consultations shall give special consideration to the importance of uninterrupted operation of nuclear reactors of the Party concerned with respect to the availability of nuclear energy for peaceful purposes as a means of achieving energy security. Both Parties shall take into account the potential negative consequences of such termination on the on-going contracts and projects initiated under this Agreement of significance for the respective nuclear programmes of either Party.

6. If either Party exercises its right of return pursuant to paragraph 4 of this Article, it shall, prior to the removal from the territory or from the control of the other Party, compensate promptly that Party for the fair market value thereof and for the costs incurred as a consequence of such removal. If the return of nuclear items is required, the Parties shall agree on methods and arrangements for the return of the items, the relevant quantity of the items to be returned, and the amount of compensation that

would have to be paid by the Party exercising the right to the other Party.

7. Prior to return of nuclear items, the Parties shall satisfy themselves that full safety, radiological and physical protection measures have been ensured in accordance with their existing national regulations and that the transfers pose no unreasonable risk to either Party, countries through which the nuclear items may transit and to the global environment and are in accordance with existing international regulations.

8. The Party seeking the return of nuclear items shall ensure that the timing, methods and arrangements for return of nuclear items are in accordance with paragraphs 5, 6 and 7. Accordingly, the consultations between the Parties shall address mutual commitments as contained in Article 5.6. It is not the purpose of the provisions of this Article regarding cessation of cooperation and right of return to derogate from the rights of the Parties under Article 5.6.

9. The arrangements and procedures concluded pursuant to Article 6(iii) shall be subject to suspension by either Party in exceptional circumstances, as defined by the Parties, after consultations have been held between the Parties aimed at reaching mutually acceptable resolution of outstanding issues, while taking into account the effects of such suspension on other aspects of cooperation under this Agreement.

ARTICLE 15 - SETTLEMENT OF DISPUTES

Any dispute concerning the interpretation or implementation of the provisions of this Agreement shall be promptly negotiated by the Parties with a view to resolving that dispute.

ARTICLE 16 - ENTRY INTO FORCE AND DURATION

1. This Agreement shall enter into force on the date on which the Parties exchange diplomatic notes informing each other that they have completed all applicable requirements for its entry into force.
2. This Agreement shall remain in force for a period of 40 years. It shall continue in force thereafter for additional periods of 10 years each. Each Party may, by giving 6 months written notice to the other Party, terminate this Agreement at the end of the initial 40 year period or at the end of any subsequent 10 year period.
3. Notwithstanding the termination or expiration of this Agreement or withdrawal of a Party from this Agreement, Articles 5.6(c), 6, 7, 8, 9, 10 and 15 shall continue in effect so long as any nuclear material, non-nuclear material, by-product material, equipment or components subject to these articles remains in the territory of the Party concerned or under its jurisdiction or control anywhere, or until such time as the Parties agree that such nuclear material is no longer usable for any nuclear activity relevant from the point of view of safeguards.
4. This Agreement shall be implemented in good faith and in accordance with the principles of international law.
5. The Parties may consult, at the request of either Party, on possible amendments to this Agreement. This Agreement may be amended if the Parties so agree. Any amendment shall enter into force on the date on which the Parties exchange diplomatic notes informing each other that their respective internal legal procedures necessary for the entry into force have been completed.

ARTICLE 17 - ADMINISTRATIVE ARRANGEMENT

1. The appropriate authorities of the Parties shall establish an Administrative Arrangement in order to provide for the effective implementation of the provisions of this Agreement.
2. The principles of fungibility and equivalence shall apply to nuclear material and non-nuclear material subject to this Agreement. Detailed provisions for applying these principles shall be set forth in the Administrative Arrangement.
3. The Administrative Arrangement established pursuant to this Article may be amended by agreement of the appropriate authorities of the Parties.

IN WITNESS WHEREOF the undersigned, being duly authorized, have signed this Agreement.

DONE at ...., this .... day of ...., 200, in duplicate.

For The Government of The United States of America	For The Governemnt of India

AGREED MINUTE

During the negotiation of the Agreement for Cooperation Between the Government of the United States of America and the Government of India Concerning Peaceful Uses of Nuclear Energy ("the Agreement") signed today, the following understandings, which shall be an integral part of the Agreement, were reached.

Proportionality

For the purposes of implementing the rights specified in Articles 6 and 7 of the Agreement with respect to special fissionable material and by-product material produced through the use of nuclear material and non-nuclear material, respectively, transferred pursuant to the Agreement and not used in or produced through the use of equipment transferred pursuant to the Agreement, such rights shall in practice be applied to that proportion of special fissionable material and by-product material produced that represents the ratio of transferred nuclear material and non-nuclear material, respectively, used in the production of the special fissionable material and by-product material to the total amount of nuclear material and non-nuclear material so used, and similarly for subsequent generations.

By-product material

The Parties agree that reporting and exchanges of information on by-product material subject to the Agreement will be limited to the following:

(1) Both Parties would comply with the provisions as contained in the IAEA document GOV/1999/19/ Rev.2, with regard to by-product material subject to the Agreement.

(2) With regard to tritium subject to the Agreement, the Parties will exchange annually information pertaining to its disposition for peaceful purposes

consistent with Article 9 of this Agreement.

For The Government of The United States of America	For The Governemnt of India

Bibliography

PRIMARY SOURCES

Agreement between Government of India and the International Atomic Energy Agency for the Application of Safeguards to Civilian Nuclear Facilities.

Agreement for Cooperation between the Government of India and the Government of the United States of America Concerning Peaceful Uses of Nuclear Energy.

Energy Statistics 2013, Report of Ministry of Statistics and Programme implementation.

GOV/1621 of 20 August 1973.

Hearing of the House Committee on International Relations on the U.S.-India Global Partnership.

Hearing of the US Senate Committee on Foreign Relations on U.S.-India Atomic Cooperation, April 28, 2006.

Henry J. Hyde United States-India Peaceful Atomic Energy Act of 2006

Implementation of the India-United States Joint Statement of July 18, 2005: India's Separation Plan.

India-US Joint Statement of March 2, 2006.

Indo-U.S. Joint Statement of 18 July 2005.

Information Circular 254/ Rev. 8/ Part II, June 2010.

Information Circular 254/Rev. 11/ Part I, 12 November, 2012.

Lok Sabha Debates, August 23, 2006, XXII (20).

Lok Sabha Debates, August 3, 2005, XI (8).

Lok Sabha Debates, December 18, 2006, XXIV (19).

Lok Sabha Debates, February 27, 2006, XVII (8).

Opening Statement of Senator Joseph R. Biden before Senate Foreign Relations Committee on "U.S-India Atomic Energy Cooperation," April 5, 2006.

Statement by Prime Minster Dr. Manmohan Singh in Parliament on his Visit to the United States, July 29, 2005.

Statement of Condoleezza Rice, US Secretary of State before the House International Relations Committee on "U.S.-India Civil Nuclear Cooperation Agreement," April 5, 2006.

Statement of Condoleezza Rice, US Secretary of State before the Senate Foreign Relations Committee on "The U.S.-India Civil Nuclear Cooperation Agreement," April 5, 2006.

Statement of John C. Rood, Acting Under Secretary of State for Arms Control and International Security before Senate Foreign Relations Committee on "The U.S.-India Civil Nuclear Cooperation Initiative," September 18, 2008.

Statement of Nicholas Burns, Under Secretary of State for Political Affairs before the Senate Foreign Relations Committee on "the U.S. and India," November 2, 2005.

Statement of Prime Minister Dr. Manmohan Singh in Rajya Sabha on the India-US Nuclear Agreement, August 17, 2006.

Statement of William J. Burns, Under Secretary of State for Political Affairs before the Senate Foreign Relation Committee on "The U.S.-India Civil Nuclear Cooperation Initiative," September 18, 2008.

Suo-Motu Statement by Prime Minister Dr. Manmohan Singh in Parliament on Civil Nuclear Energy Cooperation with United States, February 27, 2006.

Suo-Motu Statement by Prime Minister Dr. Manmohan Singh on Discussions on Civil Nuclear Energy Cooperation with the US: Implementation of India's Separation Plan, March 7, 2006.

Testimony of Ashley J. Tellis, Senior Associate, Carnegie Endowment for the International Peace before the Senate Foreign Relations Committee on "U.S.-India Atomic Energy Cooperation: Strategic and Non-Proliferation Implications," April 26, 2006.

Testimony of Ashton B. Carter, Co-Director, Preventive Defence Project, Kennedy School of Government, Harvard University on "The India Deal: Looking at the Big Picture," November 2, 2005.

Testimony of Gary Milhollin, Director, Wisconsin Project on Nuclear Arms Control before the Senate Committee on Foreign Relations, April 26, 2006.

Testimony of Michael Krepon, Co-founder, the Henry L. Stimson Centre before the Senate Foreign Relations Committee, November 3, 2005.

Testimony of Robert J. Einhorn, Senior Adviser, Center for Strategic and International Studies before the Senate Foreign Relations Committee on the "U.S.-India Civil Nuclear Deal," April 26, 2006.

Testimony of Robert L. Gallucci, Dean of the Edmund A. Walsh School of Foreign Service, Georgetown University before the Senate Foreign Relations Committee on " The Proposed US-India Nuclear

Deal," April 26, 2006.

Testimony of Ronald F. Lehman II before the Senate Foreign Relations Committee on "U.S.-Indian Nuclear Energy Cooperation: Security and Non-proliferation," November 2, 2005.

Text of the India-related waiver by the Nuclear Suppliers' Group (NSG).

SECONDARY SOURCES

Books and Journals

Andres, Gabriel Eugenio. May - December 2008, "The International Atomic Energy Agency's Safeguards System," *International Law Review*, 11

Banarjee, Dipanker. May 16-22, 1998, "Budha's Smile and National Security," *Economic and Political Weekly*, 33(20).

Basrur, Rajesh. March 2000, "Nuclear Weapons and Indian Nuclear Strategic Culture," *Journal of Peace Research*, 38(2).

__________ and Bommakanti Kartik. March 2011, "The India-China Nuclear Relationship," *Strategic Analysis*, 35(2).

__________.2006, *Minimum Deterrence and India's Nuclear Security*, California: Stanford University Press,

Bhatia, Shyam. 1979, *India's Nuclear Bomb*, Ghaziabad: Vikas Publishing House.

Bidwai, Praful and Achin Vanaik. 1999, *South Asia on A Short Fuse*, New Delhi: Oxford University Press.

____________________. 2000, *New Nukes: India, Pakistan and Global Disarmament*, United Kingdom: Signal Books.

Bidwai, Praful. July 30 - August 5, 2005, "A Deplorable

Nuclear Bargain," *Economic and Political Weekly*, 40(31).

Bukhari, Syed Shahid Hussain. January 2011, "India-United States Strategic Partnership: Implications for Pakistan," *Barkley Journal of Social Sciences*, 1(1).

Chakma, Bhumitra. February 2005, "Towards Pokhran II: Explaining India's Nuclearisation Process," *Modern Asian Studies*, 39(1).

Chari, P. R. Fall-Winter 2000, "India's Nuclear Doctrine: Confused Ambitions," The *Non-Proliferation Review*.

Chaudhury, Rahul Roy. May 2009, "India's Nuclear Doctrine: A Critical Analysis," *Strategic Analysis*, 33(3).

Chenoy, Kamal Mitra and Anuradha M. Chenoy. September 1-7, 2007, "India's Foreign Policy Shifts and the Calculus of Power," *Economic and Political weekly*, 42(35).

Dougherty, James E. January 1966, "The Non-Proliferation Treaty," *Russian Review*, 25(1)

Fani, Muhammad Ishaque. "The Indo-US Strategic Partnership in Post 9/11: Implications for Pakistan," *Pakistan Vision*, 10(2).

Firmage, Edwin Brown. October 1969, "The Treaty on the Non-Proliferation of Nuclear Weapons," *The American Journal of International Law*, 63(4)

Fitzpatrick, Mark. 2008, "US-India Nuclear Cooperation Accord: Implications for Non-Proliferation Regime," *Asia Pacific Review*, 15(1).

Freedman, Allison Marz. "USINAPC and U.S-India Nuclear Deal: Lasting Influence or One Short victory?," *CUREJ College Undergraduate Research Electronic Journal*.

Frey, Karsten. 2006, *India's Nuclear Bomb and National*

Security, New York: Routledge.

Ganguly, Sumit. Spring 1999, "India's Path to Pokhran II: The Prospects and Sources of New Delhi's Nuclear Weapons Programme," *International Security*, 23(4).

Garg, P. 2012, "Energy Scenario and Vision 2020 in India," *Journal of Sustainable Energy and Environment.*

Gopalakrishnan, A. August 27 - September 2, 2005, "Some Concerns on Indo-US Deal," *Economic and Political Weekly*, 40(35).

Goswami, Namrata. July - September, 2006, "The Essence of South-Asian Nuclear Debate," *Strategic Analysis*, 30(3).

Gupta, K. R. 2007, "Indo-US Nuclear Cooperation" in Rahul Bhonsle et al (eds.), *Indo-US Nuclear Deal* ,vol. II, New Delhi: Atlantic Publishers.

Hosur, Prashant. Spring 2010, "The Indo-US Civilian Nuclear Agreement: What's the Big Deal," *International Journal.*

Hu, Weixing. Summer 2000, "New Delhi's Nuclear Bomb: A Systemic Analysis," *World Affairs*, 163(1).

"India- United States Sign Historic Civil Nuclear Agreement," November 2008, *India Review*, 4(11).

Iyengar, P. K. 2009, "The Indo-US 'Unclear Deal'," in P. K. Iyengar et al, *Strategic Sell-Out: Indian-US Nuclear Deal*, New Delhi: Pentagon Press.

Jain, Neeraj. 2012, *Nuclear Energy: A Technology From Hell*, New Delhi: Aakar Books.

Jain, S. K. November 2-3, 2004, "Nuclear Power in India – The Fourth Revolution," *An International Journal of Nuclear power*, 18(2-3).

Jain, Sandhya. September 13, 2008, "Manmohan Outsources Political Sovereignty," *Mainstream*, XLVI (39).

Jaishankar, Dhruva. 2009, "Chronicle of a Deal Foretold: Washington's Perspective on Negotiating the Indo-U.S. Nuclear Deal" in P. R. Chari (ed.), *Indo-U.S. civil Nuclear Deal: Seeking Synergy in Bilateralism*, New Delhi: Routledge.

Jayaprakash, N. D. February 12-18, 2000, "Nuclear Disarmament and India," *Economic and Political weekly*, 35(7).

Kamath, P. M. 2009, *India's Policy of No First Use of Nuclear Weapons*, New Delhi: Anamika Publishers.

Kapur, Ashok. November 1974, "India's Nuclear Presence," *The World Today*, 30 (11).

Kapur, S. Paul and Sumit Ganguly. 2007, "The Transformation of US-India Relations: An Explanation of Rapprochement and Prospects for Future," *Asian Survey*, 47(4).

Karat, Prakash. July - September 2007, "The Indo-US nuclear Deal: Struggle to Defend National Security," *The Marxist*, XXIII (3).

__________. 2007, *Subordinate Allay: The Nuclear Deal and India-US Strategic Relations*, New Delhi: Left Word Books.

Karnad, Bharat. 2008, *India's Nuclear Policy*, London: Praeger Security International.

Kazi, Reshmi. September 2009, "India and Nuclear Testing," *Strategic Analysis*, 33(5).

Khan, Muhammad Tayyab. "The Geopolitical Dimensions of Indo-US Nuclear Deal," *Policy Perspectives*, 3(2).

Khan, Rashid Ahmad. Winter 2007, "India-US Nuclear Deal: Implications for the Region," *IPRI Journal*, VII (1).

Khana, V. N. 2000, *India's Nuclear Doctrine*, New Delhi: Samskriti.

Kumar, Ashwin. December 31, 2005 – January 6, 2006, "Indo-US Nuclear Deal: Separation of Civilian and Military Facilities," *Economic and Political weekly*, 40(53).

Lavoy, P. R. January 1, 2007, "Indian in 2006: A New Emphasis on Engagement," *Asian Survey*, 47(1).

Mathew, Thomas. Summer 2008, "Sign of Emerging Third Leg: Strengthening of India's Triad," *Journal of Defence Studies*, 2(1).

Mian Zia and M.V. Ramana. August 27- 2 September, 2005, "Feeding the Nuclear Fire," *Economic and political weekly*, 40(35).

Mishra, S. K. (ed.). 2006, *India's Nuclear Policy, Disarmament and International Security*, New Delhi: Radha Publications,

Mistry, Dinshaw. September/October 2006, "Diplomacy, Domestic Politics and U.S.-India Nuclear Agreement," *Asian survey*, 46(5).

Mohan, C. Raja. 2006, *Impossible Allies: Nuclear India, United States and the Global order*, New Delhi: Indian Research Press.

Mohite, Dilip H. and Sumita Pawar. 2001, *Pokhran Vs Chhagai Hills: The Nuclear Dilemma for India and Pakistan*, Delhi: Kalinga Publications.

Nanda, Ravi. 1998, *Strategic Compulsions of Nuclear India*, New Delhi: Lancer Books.

Nautiyal, Annpurna. September 2008, "Indo-US Civil Nuclear Deal: What is There," *Strategic Insights*, VII (4).

Noorani, A G. April 1981, "Indo-US Nuclear Relations," *Asian Survey*, 21(4).

____________. July 1967, "India's Quest for Nuclear Guarantee," *Asian survey*, 7(7).

Ollapally, Deepa M. November - December 2001,"Mixed Motives in India's Search for Nuclear Status," *Asian Survey,* 41(6).

Paddock, Carl. 2009, *India-US Nuclear Deal: Prospects and Implications*, New Delhi: Epitome Books, pp. 138-139.

Pant, Harsh V. 2011, *The US-India Nuclear Pact: Policy Process and Great Power Politics*, New Delhi: Oxford University Press.

Paranjpe, Shrikant. Summer 1986, "India's Nuclear Status: A Policy Note," *Asian Affairs*, 13(2).

Perkovich, George. 2000, *India's Nuclear Bomb: The Impact on Global Proliferation,* New Delhi: Oxford University Press.

Power, Paul F. June 1979, "The Indo-American Nuclear Controversy," *Asian Survey*, 19(9).

Rai, Ajai K. 2009, *India's Nuclear Diplomacy After Pokhran II*, Delhi: Darling Kindersley.

Raju, Suvrat. January 2010, "India's Atomic Energy Programme: Claims and Reality," *Aspects of India's Economy*, (48).

Ramana, M.V. 2012, *The Power of Promise: Examining Nuclear Energy in India, New Delhi: Penguin.*

____________. and R. C. Rammanohar Reddy (eds.). 2003, *Prisoners of Nuclear Dream*, New Delhi: Orient Longman.

Rani, Sudesh. 2010, "Indo-US Nuclear Agreement: A Critical Study," *Punjab Journal of Politics*, XXXIV (1-2).

Rhodes, Richard and Dennis Beller. January/February 2000, "The Need for Nuclear Power," *Foreign Affairs*, 79(1).

Sagan, Scot D. Winter 1996-97, "Why Do States Build

Nuclear Weapons: Three Models in Search of A Bomb," *International Security*, 21(3).

Singh, B .B. 2007, "The Hyde Act 2006: India's Nuclear Dilemma," *Atoms for Peace: An International Journal*, 1(4).

Singh, Jasjit et al. October - December 2007, "Indo-US Nuclear Deal: A Debate," *Vikalpa*, 32 (4).

Smith, Gretchen and Rekha Chakravarthi. July 26, 2008, "India-Specific Safeguards Agreement: India and American Responses," *Mainstream*, XLVI (2).

Srinivasan, M. R. May 25, 2006, "Indo-US Civil Nuclear Agreement: Expectations and Concerns," *Current Science*, 90(10).

Subramanian, R. 1975, "Nuclear India and the NPT: Prospects for Future," *Instant Research on Peace and Violence*, 5(1).

Warburg, Gerald Felix. November 5, 2012, "Non-Proliferation Policy Crossroads," *The Non-Proliferation Review*, 19(3).

Weiss, Leonard. March 2010, "India and the NPT," *Strategic Analysis*, 34(2),

____________. November 2007, U.S.-India Nuclear Cooperation: Better Later Than Sooner," *The Non-Proliferation Review*, 14(3).

Magazines, Newspapers and Issue Briefs

"A unique Development Says, Pranab," 2008, *The Hindu*, 7 September.

Akhlaque, Qudssia. 2006, "Islamabad not Fully Informed on N- Deal," *Dawn*, 11 April.

Bidwai, Praful. May 24 - June 6, 2008, "Saying No to Nukes," *Frontline*, 25(11).

____________. March 11-24, 2006, "Nuclear Deal at What Price?," *Frontline*, 23(5).

___________. September 24 - October 7, 2005, "India's Nuclear Albatross," *Frontline*, 25(20).

"BJP Questions on IAEA Draft," 2008, *The Hindu,* 12 July.

Chari, P. R. March 2006, "Parsing the Separation Plan: The Indo-US Subsidiary Deal," *IPCS Issue Brief*, no.36.

Chellaney, Brahma. 2007, "123: Text and Context, Pact Unequal, US can Scrap it at Will," *Deccan Chronicle*, 4 August.

_______________. 2007, "Questions Raised by the 123 Agreement," *The Hindu*, 14 September.

Cherian, John. March 10, 2006, "On a Mission to India," *Frontline*, 23(4).

__________. February 25 - March 10, 2006, "The China Factor," *Frontline*, 23(4).

__________. September 27 - October10, 2008, "The Morning After," *Frontline*, 25(20).

"Concerns Expressed at Indo-US Deal: Deterrence Satisfactory: NCA," 2006, *Dawn*, 13 April.

"Condition of the Hyde Act Remain: Karat," 2008, *The Hindu*, 7 September.

"Deal Will Trigger New Crises, says Iran," 2008, *The Hindu*, 6 October.

Dikshit, Sandeep. 2008, "Hyde Act is only a Domestic Legislation, Says Boucher," *The Hindu*, 5 March.

Dikshit, Sandeep. 2008, "Nuclear Energy is Cover, Intent is Strategic US Ties: CPI (M)," *The Hindu*, 14 May.

____________. 2011, "Russia Alone Proposed a Way out of Tightened NSG Rules, Says Envoy", *The Hindu*, 8 December.

"Executive Cannot Act Independently on Parliament on Nuclear Deal," 2007, A statement issued by V. R. Krishna Iyengar, justice P.B. Sawant and Justice H.

Suresh, *The Hindu*, 13 September.

Ghose, Arundhati. 2007, "123 Agreement: No Threat to India's Strategic Programme," *The Tribune*, 12 November.

Goldschmidt, Pierre. 1999, "The IAEA Safeguards System Moves into 21st Century," *IAEA Bulletin*, 41(4).

____________. 2001, "Strengthened Safeguards: Meeting Present and Future Challenges," *IAEA Bulletin*, 43(4).

"Good Day for India, Bad Day for Non-proliferation: Strobe Talbot," 2005, *Daily Times*, 25 July.

Hibbs, Mark. April 27, 2010, "Pakistan Deal Signals China's Growing Nuclear Assertiveness," *Nuclear Energy Brief*, Carnegie Endowment for International Peace.

Horner, Daniel. April 2011, "Chinese-Pakistani Reactor Deal Moves Ahead," *Arms Control Today*.

"Hyde Act and Nuclear Scientists Note," 2006, *The Hindu*, 6 December.

"IAEA Safeguards Agreement is Against India's Interests," 26 July, 2008, The Left parties Statement on IAEA Safeguards Agreement, *People's Democracy*, XXXI (28).

"India Wants NSG Partners to Fully Honour their Commitments," 2011, *The Hindu*, 11 August.

"Indo-US Deal Getting Bipartisan Support," 2006, *Dawn*, 7 April.

Isola, Justine. August 26, 2010, "If India Tests? The Implications for India-US Civil Nuclear Deal," *IDSA Issue Brief*.

Iyengar, P. K and M. Gupta. 2006, "India-US Deal Will Destroy Nuclear Research," *The Asian Age*, 5 April.

________. 2009, "Pokhran-II Revisited," *Business Daily*, 18 December.

Kalam, A. P. J. Abdul and Srijan Pal Singh. 2011, "Nuclear Power is our Gateway to a Prosperous Future," *The Hindu,* 6 November.

Karnad, Bharat. 2005, "Remember the Tritium," *The Asian Age,* 7 August.

____________. 2006, "Parliament Must Examine PM's Nuclear Deal," *The Asian Age,* 14 May.

____________. 2006, "Turning India into Nuclear Cripple," *The Asian Age,* 7 April.

____________. 2006, "Why the Nuclear Deal is A Disaster," *The Asian Age,* 18 March.

Krepon, Michael. 2011, "Another Contentious Nuclear Issue," *The Hindu, 12 July.*

Krishnan, Ananth. 2013, "China-Pakistan Signed Deal on 1000 MW Reactor," *The Hindu,* 26 March

Kumar, Vinay. 2008, "Nothing Expected in IAEA Safeguards, Karat," *The Hindu,* 12 July.

____________.2008, "Nothing Unexpected in IAEA Safeguards Agreement: Karat," *The Hindu,* 12 July.

Lafranchi, Howard. 2008, "Fallout of US-India Nuke Deal," *The Christian Science Monitor,* 24 October.

____________. 2010, "US Objects to China-Pakistan Nuclear Deal Hypocritical?" *The Christian Science Monitor,* 16 June.

"Left Told Deal Will Not be Pursued: Advani," 2008, *The Hindu,* 5 March.

Malik, Ahmad Rashid. 2006, "Japan's Reaction to Indo-US N-Deal," *The Nation,* 6 March.

McGoldrick, Fred. January/February 2011, "The Road Ahead For Export Controls: Challenges for the Nuclear Suppliers Group," *Arms Control Today.*

Mehdudia, Sujay. 2013, "India Poised to Become Global Solar Power: World Bank", *The Hindu,* 12 December.

Merchant, Minhaz. 2012, "Powerless in India," *The Times of India*, 12 July.

"N Deal Will Hit Our Strategic Programme," 2007, Interview of Brijesh Mishra with Karan Thapar, *The Pioneer*, 9 August.

Nagesh, S. 2013, "In Nuclear Slow down Small is Better," *The Hindu*, 7 September.

"NSG Waiver Enables India to Sign Agreements with Countries: Pranab," 2008, *South Asian Times*, 6 October.

Panigrahi, Ragunath. October 2010, "India's Civil Nuclear Diplomacy," *Civil Service Chronicle*, XXI (4).

Parashar, Sachin. 2011, "Now France Reaffirms NSG Rules Won't Affect Ties," *The Times of India*, 2 July.

Prasad, A. N. 2006, "Down to the Last Atom," 27 February, *Outlook*.

Priest, Jan. 1995, "IAEA Safeguards and NPT: Examining Interconnections," *IAEA Bulletin* I.

Purkayastha, Prabir. 2008, "Does the IAEA Agreement Hide Us from Hyde Act?" *The Hindu*, 14 July.

Radyuhin, Vladimir. 2011, "New ENR Norms Won't Hit Waiver to India: Russia," *The Hindu, 16* July.

_______________. 2011, "Russia Silent on Whether NSG Ban will Cover India," *The Hindu*, 30 June.

_______________. 2011, "Twenty Five Years After Chernobyl," *The Hindu*, 26 April.

Raju, Suvrat and M.V. Ramana. 2011, "Why Kundankulam is Untenable," *The Hindu*, 12 November.

Ramachandran, R. August 30 - September 12, 2011, "On Slippery Ground," *Frontline*, 28(16).

_______________. December 16-29, 2006, "Shifting of Goalposts," *Frontline*, 23(25)

______________. October 25 - November 7, 2008, "Waiting Traps," *Frontline*, 25(22).

______________. September 27 - October 10, 2008, "Nuclear Hurdles," *Frontline*, 25(20).

______________. August 16-29, 2008, "Inching Closer," *Frontline*, 25(17)

______________. September 27 - October 10, 2008, "The Truth, yet Again," *Frontline*, 25(20).

Ramana, M.V. 2003, "India's Nuclear Command Authority," *Daily Times*, 9 January.

Ramdas, L. July 4-17, 1998, "Pokhran and its Fall Out," *Frontline*, 15(14).

Ramesh, M. 2013, "Hydro: Ambitions Watered Down", *The Hindu Business Line*, 1 September.

Santhanam, K. and Ashok Parthasarathi. 2009, "Pokhran-II Thermonuclear Test, a Failure," *The Hindu*, 17 September.

Shourie, Arun. 2006, "Now Let the PM Square this Circle", *The Indian Express*, 28 November.

___________. 2006, "Facts Versus Government's Fiction," *The Indian Express*, 22 December.

___________. 2006, "Time to Deal with Aftermath," *The Indian Express*, 30 November.

___________. 2008, "An Empty Claim," *The Indian Express*, 8 September.

Shuja, Sharif. 2007, "Pakistan Feels Jilted By US-India Nuclear Deal," *News Weekly*, 17 February.

Srinivasan, M. R. 2006, "India May Lose Control of Its Nuclear Future," *The Hindu*, 14 December.

Subba Rao, Buddhi Kota. June 6-19, 1998, "The H Bomb Issue is Crucial," *Frontline*, 15(12).

Subramanian, T.B. 2008, "A Breakthrough, Says Kakodkar," *The Hindu*, 8 September.

Telis, Ashley J. July 16, 2010, "The China-Pakistan Nuclear Deal: Separating Fact from Fiction," *Policy Outlook*, Carnegie Endowment for International Peace.

Telis, Ashley J. 2005, "Should US Sell Nuclear Technology to India?-Part-II," *Yale Glob*al,

"Tighten Draft Waiver for India, NSG Urged," 2008, *The Hindu*, 16 August.

Varadarajan Siddharth.2008, "Bush Signing Statement on 123 Leaves Flaws Intact," *The Hindu*, 10 October.

__________________. 2006, "Was Bush Speech A Warning on Separation," *The Hindu*, 1 March

__________________. 2008, "IAEA Safeguards Grant India Nuclear Weapon Status," *The Hindu*, 24 July.

__________________. 2008, "Most IAEA Members Worried that Precedent Being Set for Pakistan", *The Hindu*, 1August.

__________________. 2008, "New NSG Draft Has Small but Significant Changes," *The Hindu*, 5 September.

__________________. 2008, "Perpetuity of Safeguards Only with Perpetuity of Fuel Supply," *The Hindu*, 13 July.

__________________. 2008, "Waiver Enables Member States to Provide India Full Civil Nuclear Cooperation," *The Hindu*, 7 September.

__________________. 2005, "Nuclear Bargain May Prove Costly in the Long Run," *The Hindu, 20* July.

__________________. 2007, "A Guide to the Provisions of 123 Agreement," *The Hindu*, 4 August.

__________________. 2007, "Deal Breather not Deal Breaker," *The Hindu*, 20 August.

__________________. 2007, "Kakodkar Supports Nuclear Deal," *The Hindu*, 28 July.

__________________. 2007, "Nuclear Deal Satisfactory:

Kakodkar," *The Hindu,* 28 August.

____________________. 2008, "India Reiterates 'Legally Binding' Nature of 123 Pact's Provisions," *The Hindu,* 12 October.

____________________. 2009, "India and the Additional Protocol," *The Hindu,* 7 March.

Vyas, Neena. 2007, "123 Agreement an Assault on Nuclear Sovereignty: BJP," *The Hindu,* 5 August.

__________. 2008, "India has Walked into a Trap: BJP," *The Hindu,* 7 September.

Yadav, Basudeo Prasad. October 2008, "Indo-US Civil Nuclear Cooperation Agreement," *Pratiyogta Darpan.*

Reports

Bharadwaj, Anshu et al. "Nuclear Power in India: The Road Ahead," *SCTEP Technical Report*, Center for Science, Technology and Policy.

"Development of Nuclear Energy Sector in India," *IDSA Taskforce Report No. 201.*

Gupta, Sonika et al. 2007, "Assessing the Indo-US Deal on Civil Nuclear Cooperation: Managing Risks and Opportunities," *NIAS Report.*

Gupta, Sonika et al. September 2006, "Assessing the Indo-US Nuclear Deal on Civil Nuclear Cooperation: Forging a New Partnership," Report of National Institute of Advanced Studies.

Jacob, Jabin T. March 2006, "Indo-US Civil Nuclear Deal: The China Factor," *IPCS Special Report 14*, Institute of Peace and Conflict Studies.

Kerr, Paul K. December 15, 2011, "US Nuclear Cooperation with India: Issues for Congress," *CRS Report for Congress*, Congressional Research Service.

Kronstadt, K. Alan. August 12, 2008, "India-US Relations," *CRS Report for Congress,* Congressional Research Service.

Kronstadt, K. Alan. September 8, 2005, "US-India Bilateral Agreements in 2005," *CRS Report For Congress,* Congressional Research Service.

Manjunath, K.S. and Beryl Anand. February 2006, "India's Credible Minimum Deterrence," *IPCS Special Report 13,* Institute of Peace and Conflict Studies.

Mian, Zia et al. March 2007, "Plutonium Production in India and the US-India Nuclear Deal," in Henry Sokolski, *Gauging US-India Strategic Cooperation,* Report published by Strategic Studies Institute.

Mian, Zia et al. September 2006, "Fissile Materials in South Asia: The Implications of US-India Nuclear Deal," A Research Report of International Panel on Fissile Materials.

Mirza, Nasrullah and M. Sadiq. January 2008, "Indo-US 123 Agreement: Impacts on Deterrence Stability in South Asia," *Research Report 07,* South Asian Strategic Stability Institute.

Ramana, M.V. March 2007, "Nuclear Power in India: Failed Past Dubious Future" in Henry Sokolski (ed.), *Gauging U.S.-India Strategic Cooperation,* Report of The Strategic Studies Institute.

Squassoni, Sharon.. December 22, 2006, "India's Separation Plan: Issues and Views," *CRS Report for Congress,* Congressional Research Service.

Sridhar, Seema et al, "Promoting Strategic and Missile Stability in Southern Asia," *IPCS Special Report 17,* Institute of Peace and Conflict Studies.

Stephenson, John and Peter Tyman. March 2007, "Will the US-India Civil Nuclear Cooperation Initiative

Light India" in Henry Sokoloski (ed.), *Gauging US-India Strategic Cooperation*, Report published by Strategic Studies Institute.

Tasleem, Sadia. 2008, "Indo-U.S. Civil Nuclear Cooperation: Altering Strategic Positioning and Balance of Power in South Asia," *RSS Policy Studies* 44.

Telis, Ashley J. 2005, "India as a New Global Power: An Action Agenda for the United States," Carnegie Endowment for International Peace.

Internet Sources

Alam, Mohammed Badrul. "India's Nuclear Doctrine: Context and Constraints." Available at http://casi.sas.upon.edu/system/files/Heidelberg+paper+-+2002.pdf, (accessed on 22 August 2011).

Albright, David and Susan Basu. April 5, 2006, "Neither a Determined Proliferator, Nor a Responsible State: India's Record Needs Scrutiny." Available at: http://www.isis-online.org/uploads/isis-reports/documents/indiacritique.pdf, (accessed on 14 November 2013).

"Atoms for Peace Speech," 1953, An Address by Mr. Dwight D. Eisenhower, President of United States of America to the 47th Plenary Meeting of the United Nations General Assembly. Available at http//www.iaea.org/about/ atoms for peace _speech. htm, (accessed on 10 October 2012).

Bajoria, Jayshree and Esther Pan. November 5, 2012, "The US-India Nuclear Deal." Available at http//www.crf.org/india/us-india-nucleardeal/19663, (accessed on 24 February 2012).

Bajpai, Kanti. "India's Nuclear Posture After Pokhran II." Available at http://www.ceri-sciencespo.com/archieve/jan01/pokhranpdf, (accessed on 26

September 2011).

Bidwai, Praful. "Nuclear Waiver Blow to Non-Proliferation." Available at http://www.ipsnews.net/print.asp?idnews=43818, (accessed on 2 March 2013).

______________. July 11, 2008, "India/US: Deceptions Surface in Nuclear Deal." Available at http://www.ipsnews.net/2008/07/india-us-deceptions-surface-in-nuclear-deal/, (accessed on 23 October 2012).

"BJP Lampoons Manmohan Singh on Proposed IAEA – India Specific Safeguards," July 14, 2008. Available at www.news.oneindia.in/2008/07114/bjp/-lampoons-manmohan-proposed-iaea-india-specific-safeguards-121090400htm, (accessed on 27 September 2012).

Boureston, Jack and Charles D. Ferguson. "Strengthening Nuclear Safeguards: Special Committee to Rescue?" Available at www.armscontrol.org/print/951, (accessed on 1 September 2012).

Bucher, R. G. "India's Baseline Plan for Nuclear Energy Self Sufficiency," January 2009. Available at http://www.ipd.anl.gov/anlpub./2010/05/67057.pdf, (accessed on 23 November 2011).

Chellaney, Brahma. August 15, 2008, "Serious Implications for India in the NSG Draft." Available at http//www.redif.com/news/2008/aug/15brahma.html, (accessed on 4 April 20102).

_________________. July 12, 2008. "The India-IAEA Safeguards Agreement Fact Sheet: Let the Facts Speak Themselves." Available at www.chellaney.net/2008/07/12/india-iaea-safeguards-agreemet-factsheet, (accessed on 13 September 2012).

Chittaranjan, Kalpana. "The IAEA as a Safeguards Regime." Available at http://www.nuclearfiles.org/menu/key-issues/nuclear-energy/history/article—chitranjan-kalpana.htm, (accessed on 22 December 2012).

Corner, Ian O. "India's Clever Nuclear Power Programme." Available at http//leadenergy.org/2011/03/indias-clever-nuclear-power-programme, (accessed on 18 November2011).

Crail, Peter. "AEA Approves India's Additional Protocol." Available at www.armscontrol.org/oct/2009_04/indiaprotocol, (accessed on 7 August 2012).

Devraj, Ramjet. March 6, 2006, "India Deal Makes US A Nuclear Proliferator." Available at www.commondreams.org/headlines06/0304-02htm, (accessed on 07 January 2014).

Goldschmidt, Bertrand. "The Origins of International Atomic Energy Agency." Available at http//www.iaea.org/publications/Magzines/Buletein/1940128121, (accessed on 1 August 2012).

"Government's Move a Shocking Betrayal: Left." Available at http://www.reiff.com/news/2008/jul/10ndeal8, (accessed on 22 September 2012).

Gulati, Sahib. "The US Domestic Debate on Indo-US Nuclear Deal Under Obama Administration: Implications for India." Available at http//www.observerindia.com/ccms/expert/orfonline/modules/analyss/attachments/implications-for-india-12393120213.pdf, (accessed on 3 April 2012).

Horner, Daniel. July - August, 2010, "NSG Makes Little Headway at Meeting." Available at http://www.arms.org/net/2010_07/NSGMtg, (accessed on 5 May 2013)

"How Many Nuclear Warheads Does India Need." Available at http//www.ipcs.org/article/nuclear/howmany-warheads-does-india-need, (accessed on 20 October 2011).

"Hyde Act, 123 Agreement and India," Excerpts of P. Chidambaram's Speech on the Motion of Confidence in the Lok Sabha. Available at: http//www.sarrijan.com/discuss/34-options/80-hyde-act-123-agreement-and-india, (accessed on 25 March 2012).

"Hydro Power Projects: Social and Environmental Perspective." Available at http://www.socialissuesindia.files.wordpress.com/2011/12/hydro-power-development-in-india.pdf, (accessed 21 January 2014).

"IAEA Safeguards Overview: Comprehensive Safeguards Agreements and Additional protocols." Available at http://www. Iaea.org/publications/Factsheets/English/sg_overview.htm, (accessed on 22 October 2013).

"IAEA Safeguards: Stemming the Spread of Nuclear Weapons." Available at http://www.iaea.org/publications/Factsheets/English/S1safeguard.pdf, (accessed on 12 August 2012).

"India's Energy Security: Key Issues impacting the India Oil and Gas sector." Available at http://www.ey.com/Publication/vwLUAssets/Indias_energy_security/$FILE/India-s_energy_security.pdf, (accessed on 11 February 2014).

"India's Missile Programme." Available at http://www.defencepk/forums/India-defence/37481-india's-missile-programme.html, (accessed on 1 November 2011).

"India's Nuclear Weapons Program: The Beginning: 1944-1960." Available at http://www.nuclearweapon archieve.org/India/Indiaorigin.htm, (accessed on 5 July 2011).

"India's Nuclear weapons Programme: India's First Bomb: 1967-1974." Available at http://www.nuclearweapon archive.org/India/Indiasfirstbomb.html/, (accessed on 7 July 2011).

"India's Nuclear Weapons Programme: On to Weapons Development: 1960-1967." Available at http://www.nuclearweaponarchive.org/India/indiaweapons development.html, (accessed on 7 July 2011).

"India's Renewable Energy Sector: Potential and Investment Opportunities." Available at www.direc2010.gov.in/pdf/India's-RE-Sector-Potential-and.Investment-opportunities-SSM.pdf, (accessed on 23 December 2013).

"India-IAEA Safeguards Agreement Worse than Useless, A Shame." July 11, 2008, Available at http:/rediff.com/news/2008/july/11ndeal2.htm, (accessed on 17 September 2012).

"Indian PM's Statement in Parliament: We Will Shut down CIRUS Reactor, Shift Apsara Reactor," March, 7, 2006. Available at http://in.rediff.com/news/2006/mar/07ndeal2.html, (accessed on 23 November 2013).

"Indo-US Nuclear Deal: Speakers Urge NSG to Adopt Non-discriminatory, Criteria Based Approach." Available at www.thenewstribune.com/2013/06/27indo-us-nuclear-deal-speakers-urge-nsg-to-adopt-non-discriminatory-criteria-based-approach, (accessed on 19 March 2014).

"International Atomic Energy Agency." Available at: http//en.wiki.org.wiki/international_Atomic_Energy_Agency, (accessed on 3 August 2012).

"Inventory of Geological Resources of Coal in India." Available at http://www.coal.nin.in/welcome.html, (accessed on 12 December 2013).

"Issues and Questions on July 18 Proposal for Nuclear cooperation with India." Available at http:// www.armscontrl.org/pdf/2005/118_India_Ltr_ congress.pdf, (accessed on 12 September 2013).

Jain, S. K. "Nuclear Power: An Alternative." Available at http//www.npcil.mic.in/pdf/nuclear%20power%20/ an%20alternatve.pdf, (accessed on 18 November 2011).

Jayal, D. "Indian Nuclear Doctrine: A Discussion." Available at http:// www. bharat.rakshak.com/monster/issue 2-3/jayal.html, (accessed on 22 September 2011).

Jyantha Dhanapala and Daryl Kimbal, July 10, 2008, "A Non-Proliferation Disaster." Available at www.carniew geendowment.org/2008/07/10/nonproliferation-disdaster/5e8271912012, (accessed on 2 October 2012) .

Kanwal, Gurmeet. "India's Nuclear Doctrine and Policy." Available at http//www.idsa.org/on-feb-1-01html, (accessed o 24 September 20011).

________________. "Indo-US Nuclear Deal and Non-Proliferation: Some Views from US." Available at http/ /www.ipcs.org/article/nuclear/indo-us-nuclear-deal-and-non-proliferation-some-views-from-2368html, (accessed on 23 October 2013).

Kaul, Aditya. June 27, 2011, "Nuclear Suppliers Group Revises Guidelines, India Wary." Available at http;// www.dnaindia.com/india/1559498/report-revises-guuidelines-india-wary, (accessed on 2 February 2013).

Kesler, Glenn. 2006, "India Nuclear Deal May face Hard Sell," *The Washington Post*, 4 April. Available at www.washingtonpost.com/wpdyun/content/article/2006/04/02/Ar2006040201315.html?referrer=email&referrer=email, (accessed on 22 April 2008).

Kimbal, Daryl G. July 26, 2006, "Impact of the Nuclear Deal on India's Fissile Material Production Capacity." Available at http://www.armscontrrl.org/print/549, (accessed on 4 April 2013)

______________. September 6, 2008, "Text, Analysis and Responses to the NSG Statement on Civil Nuclear Cooperation with India." Available at http://www.armscontroll.org/node/3345, (accessed on 7 July 2012)

______________. "Seeing Through Spin: "Critics" Rebut White House on the US-India Nuclear Cooperation Plan." Available at http://www.armscontrol.org/pressroom/2006/20060309-india-cretics-rebut-WH, (accessed on 12 October 2013).

Krepon, Michael. September 8, 2008, "Likely Consequences of the Nuclear Suppliers Group Decision." Available at http://www.stimson.or/essays/likely-consequences-of-the-nuclear-suppliers-group-decision/, (accessed on 22 November 2013)

____________. September 1, 2008, "US-India Nuclear Deal Now Done, Reckoning to Come Later." Available at http://www.stimson.org/essays/us-india-nuclear-deal-now-done-reckoning-to-come-later, (accessed on 7 November 2013).

Leslie, Russell et al. "Ensuring Effective safeguards Coverage of States with Small Quantities Protocol." Available at http://www.dfat.gov.au/asno/publications/ensuring_effective _safegaurds_coverage_state_small_quantities_protocol.pdf (accessed on 11 July 2013)

Lewis, Jeffry. September 7, 2008, "India Gets NSG Waiver." Available at http://lewis.armscontrollwork.com/archive/2008/nuclear-deal-passes-nsg, accessed on, 20 March 2013.

Mahapatra, Chintamani. March 10, 2008, "Hyde Act Controversy: Is it really pertinent." Available at http//www.sarkaritel.com/news and features/infa/march2008/10hyde_contronersy.htm, (accessed on 20 March 2012).

"Mapping India's Renewable Energy Growth potential: Status and Outlook 2013." Available at http://www.ey.com/Publication/vwLUAssets/Mapping_Indias_Renewable_Energy_growth_potential/$FILE/EY-Mapping-India's-Renewable-Energy-growth-potential.pdf, (accessed on 2 February 2014).

Mian, Zia and M.V. Ramana. "Wrong Ends, Means and Needs: Behind the US Nuclear Deal with India." Available at http://www.armscontrol.org/oct/2006/01-02/JANFEB-indiafeature#bio, (accessed on 1 September 2013).

Mohammad, Adil Sultan. July 2006, "India-US Civil Nuclear Agreement: Implications on South Asian Security Environment." Available at http//www.Stimson.org/images/uploads/research-pdfs/Adilsultan.pdf, (accessed on 22 October 2012).

"New NSG Guidelines Limit India's Access to Sensitive Nuclear Technology."Available at http:www.wise interational.org/node/3759, (accessed on 22 November 2012)

"NSG Imposes More Stringent Conditions for E and R Transfers." Available at www.nei.org/fillefolder/IssueSumaaryNSGGuidelines.pdf, (accessed on 22 April 2013).

"Nuclear Power in India." Available at: http://www.world-nuclear.org/info/Country-Profiles/Countries-G-N/India/#.UXzs8rVTAnE, accessed on (20 November2011).

"Nuclear Suppliers Group (NSG)." Available at http://cns.miis.edu/inventory/pdfs/nsg.pdf, (accessed on 12 June 2013).

Pant, Harsh V. "The US-India Nuclear Deal: The End Game Begins," January 29, 2006. Available at http://www.worldsecurityntework.com/India/Pant-Dr-Harsh-V/The-US-India-Nuclear-Deal-The-End-Game-Begins, (accessed on 27 September 2013).

"Per Capita Energy Consumption." Available at http://www.energyrealities.org/chapter/meeting-our-need/item/per-capita-energy-consumption/erp327B7c729A3B31D2B, (accessed on 22 January 2014).

Perkovich, George. "The US-India Nuclear Deal." Available at http://www.icnnd.org/Documents/US_India_Nuclear_Deal.doc, (accessed 10 November 2013).

Qadir, Shaukat. "An Outsider's View." Available at www.india-seminar.com/2006/560/560/560%20shaokat%20qadir.htm, (accessed on 18 March 2014)

"Remarks of Secretary of State Condoleezza Rice at the Senate Foreign Relations Committee on the US-India Civil Nuclear Cooperation Initiative," April 5, 2005. Available at http//www.carnegieendowment.org/static/npp/reports/rice_4-5-06.pdf, (accessed on 22 July 2012).

Samadder, Asit and Surbhi Arora. 10 September, 2012, "Natural Gas Scenario in India: An Overview," Available at: http://ssrn.com/abstract=2144061, (accessed on 12 January 2014).

Sanders, Ben. 2004, "IAEA Safeguards and the NPT." Available at http://www. unider.org/pdf/articles/pdf/art2189.pdf, (accessed on 9 December 2012).

Sauer, Tom. May 2010, "Drivers Behind the Nuclear Deal with India: US Domestic Politics or Geostrategic Concerns." Available at: www.suc.kiley-ven-be/web/files/11/72wi6-115.pdf, (accessed on 25 May 2012.

Saxena, Praveen and Arun Kumar. October, 2010, "Hydropower Development in India." Available at http://www.org/Paper2010/TSA01.pdf, (accessed on 11 February 2014).

"Securing Tomorrow's Energy Today: Policy and Regulations," February 2013. Available at https://www.deloitte.com/assets/DcomIndia/Local%20Assets/Documents/IEC%202013/Long_Term_Energy_Security.pdf, (accessed on 22 November 2013).

"The IAEA Safeguards Agreement Explained." Available at www.rediff.com/news/2008/jul/15sd1.htm, (accessed on 19 March 2014)

"The Safeguards System of International Atomic Energy Agency." Available at http://www.iaea.org/ourwork/sv/safeguards/documents/safeguardssystem.pdf, (accessed on 2 June 2012).

"US-India Nuclear Deal." Available at: http://www.reachingcriticalwill.org/resources/fact-sheets/critical-issues/5451-us-india-nuclear-deal, (accessed on 12 December 2013).

Varadarajan, Siddharth. July 16, 2008, "India's IAEA Draft: Invoking Para 29 won't be Easy." Available at www.rediff.com/news/2008/jul/16guest.htm, (accessed on 1 January 2013).

__________________. July 14, 2008, "Termination of

Safeguards on Indigenous Reactors: what Gov/1621 Really Says." Available at www.Svardarajan.blogspot.in/2008/07/termination-of-safeguards-on-indigenous.htm., (accessed on 6 September 2012).

Varma, K. J. M. March 18, 2006, "India's Special Status to Affect South Asia." Available at www.rediff.com/news/2006/news/2006/mar/17ndeal14.htm, (accessed on 19 March 2014).

Viski, Andrea. May 2012, "The Revised Nuclear Suppliers Group Guidelines: A European Union Perspective." Available at www.nonproliferation.en/documents/nonproliferation-nationpapers/andreaviski47ba1277ab879pdf, (accessed on 3 March 2013).